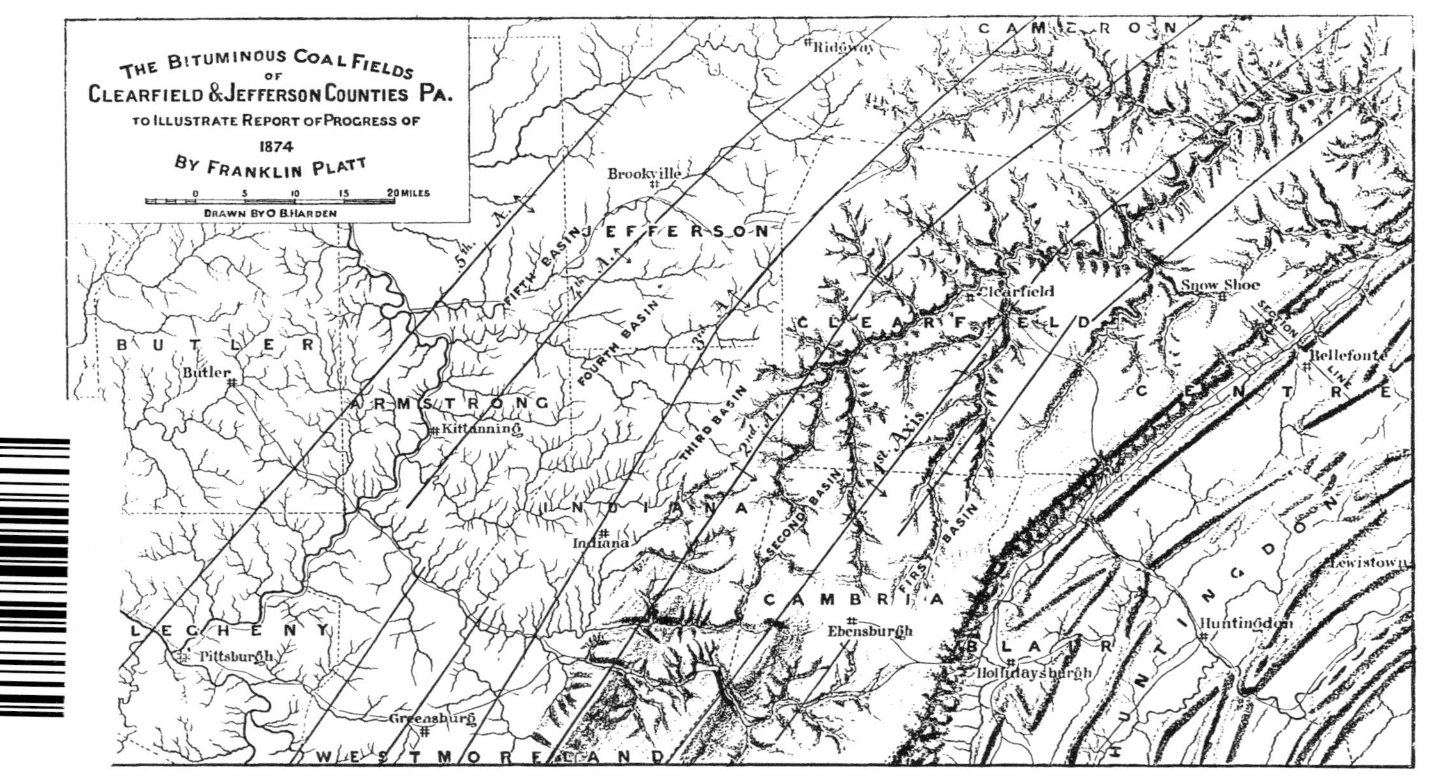
THE BITUMINOUS COAL FIELDS
OF
CLEARFIELD & JEFFERSON COUNTIES PA.
TO ILLUSTRATE REPORT OF PROGRESS OF
1874
BY FRANKLIN PLATT
0 5 10 15 20 MILES
DRAWN BY O. B. HARDEN
CAMERON
Ridgway
Brookville
JEFFERSON
FIFTH BASIN
FOURTH BASIN
THIRD BASIN
SECOND BASIN
FIRST BASIN
5th. A.
4th. A.
3rd. A.
2nd. A.
1st. Axis
Clearfield
CLEARFIELD
Snow Shoe
SECTION LINE
Bellefonte
CENTRE
BUTLER
Butler
ARMSTRONG
Kittanning
INDIANA
Indiana
CAMBRIA
Ebensburgh
BLAIR
Hollidaysburgh
HUNTINGDON
Huntingdon
Lewistown
LEGHENY
Pittsburgh
Greensburg
WESTMORELAND

SECOND GEOLOGICAL SURVEY OF PENNSYLVANIA:

1874.

REPORT OF PROGRESS

IN THE

CLEARFIELD AND JEFFERSON DISTRICT

OF THE

BITUMINOUS COAL-FIELDS

OF

WESTERN PENNSYLVANIA.

BY

FRANKLIN PLATT

ILLUSTRATED

WITH 139 WOOD-CUTS AND 10 MAPS AND SECTIONS

HARRISBURG:
PUBLISHED BY THE BOARD OF COMMISSIONERS
FOR THE SECOND GEOLOGICAL SURVEY.
1875.

Stereotyped by
Singerly Printing and Publishing House,
Harrisburg, Pa.

Printed by
B. F. MEYERS, *State Printer*,
HARRISBURG, PA.

BOARD OF COMMISSIONERS.

His Excellency, JOHN F. HARTRANFT, *Governor*,
and *ex-officio* President of the Board, Harrisburg.

ARIO PARDEE, - - - - - - Hazleton.
WILLIAM A. INGHAM, - - - - Philadelphia.
HENRY S. ECKERT, - - - - - - Reading.
HENRY M'CORMICK, - - - - - Harrisburg.
JAMES MACFARLANE, - - - - - Towanda.
JOHN B. PEARSE, - - - - - - Philadelphia.
ROBERT B. WILSON, M. D., - - - Clearfield.
Hon. DANIEL J. MORRELL, - - - Johnstown.
HENRY W. OLIVER, - - - - - Pittsburg.
SAMUEL Q. BROWN, - - - - - Pleasantville.

SECRETARY OF THE BOARD

JOHN B. PEARSE, - - - - - Philadelphia.

STATE GEOLOGIST

PETER LESLEY, - - - - - - Philadelphia.

PHILADELPHIA, *December* 31, 1874.

Prof. J. P. LESLEY,

State Geologist:

SIR:—I have the honor to submit the following detailed report of the operations of my party in the Clearfield and Jefferson District of the Bituminous Coal Regions of Pennsylvania, during the season of 1874.

Although field work was begun at the earliest possible moment after receiving my appointment as Assistant, and accompanying instructions (July 1, 1874), it was July 9th when I reached Clearfield county, July 17th when Mr. R. H. Sanders, as Senior Aid, reported to me for duty, and August 8th when Messrs. H. J. Fagen and C. A. Young, Aids, reported to me at Osceola, Clearfield county, completing the organization of my party.

Work was prosecuted during July and part of August in Clearfield and Centre counties; and in the latter part of August, in September and October in Jefferson and Clarion counties. In November a partial examination was made of the region around Marion, in Forest county, of the Potato creek region, in M'Kean county, and of the Fall Brook region, in Tioga county. The examination of the entire line of the Bennett's Branch (Low Grade R. R.) division of the Allegheny Valley railroad was about completed when the heavy snows put an end to field work.

On the 21st of October, Messrs. Sanders, Young and Fagen were detached from work in the Bituminous Coal Measures proper, and instructed to proceed to Bellefonte, in Centre county, to complete maps of the Bellefonte gap, and to make a cross section from the Nittany mountain to the crest of the Allegheny mountain. This was successfully accomplished in November, and since that time Mr. Sanders has assisted me in office work.

Messrs. Young and Fagen remained in Bellefonte and its vicinity during the greater part of December, to secure as complete a suite of specimens of the Silurian, Devonian and Lower Carboniferous measures as might be possible, for the cabinet of the Survey at Harrisburg.

The coal of the measures below the Great Conglomerate was also examined by Mr. Young, at the opening made on it on the eastern escarpment of the Allegheny mountain, near Tipton station, Blair county.

During the winter months, Messrs. Sanders, Young and Fagen have been actively engaged assisting me in the office work connected with the field work of the past season.

While so many persons have voluntarily assisted the field work of the survey it seems invidious to select a few names for mention; and yet it is only right that I should especially note the help received from Mr. L. J. Lingle, of Osceola, Clearfield county; Dr. R. V. Wilson, at Clearfield; Mr. J. L. Sommerville, of Bellefonte, Centre county; Mr. P. W. Jenks, of Punxatawney, Jefferson county; Capt. Brinker, of Fairmount, Clarion county; Mr. Lyman, of Roulette, Potter county; and Mr. John A. Wilson, Civil Engineer, of Philadelphia.

Messrs. Sanders, Young and Fagen have rendered zealous and efficient service.

I cannot conclude without returning to you my sincere thanks for the promptness with which you have furnished to me all the assistance needed for a vigorous prosecution of the work entrusted to my charge. I remain, with much respect

Your obedient servant,

FRANKLIN PLATT,

Assistant Geologist.

TABLE OF CONTENTS

OF THE SEVERAL CHAPTERS.

APPENDIX.

Extracts from H. D. Rogers' Memoir on the Coal Formation of Pennsylvania, on pages 796–810, *of the Final Report of the Geology of Pennsylvania, published in* 1858.

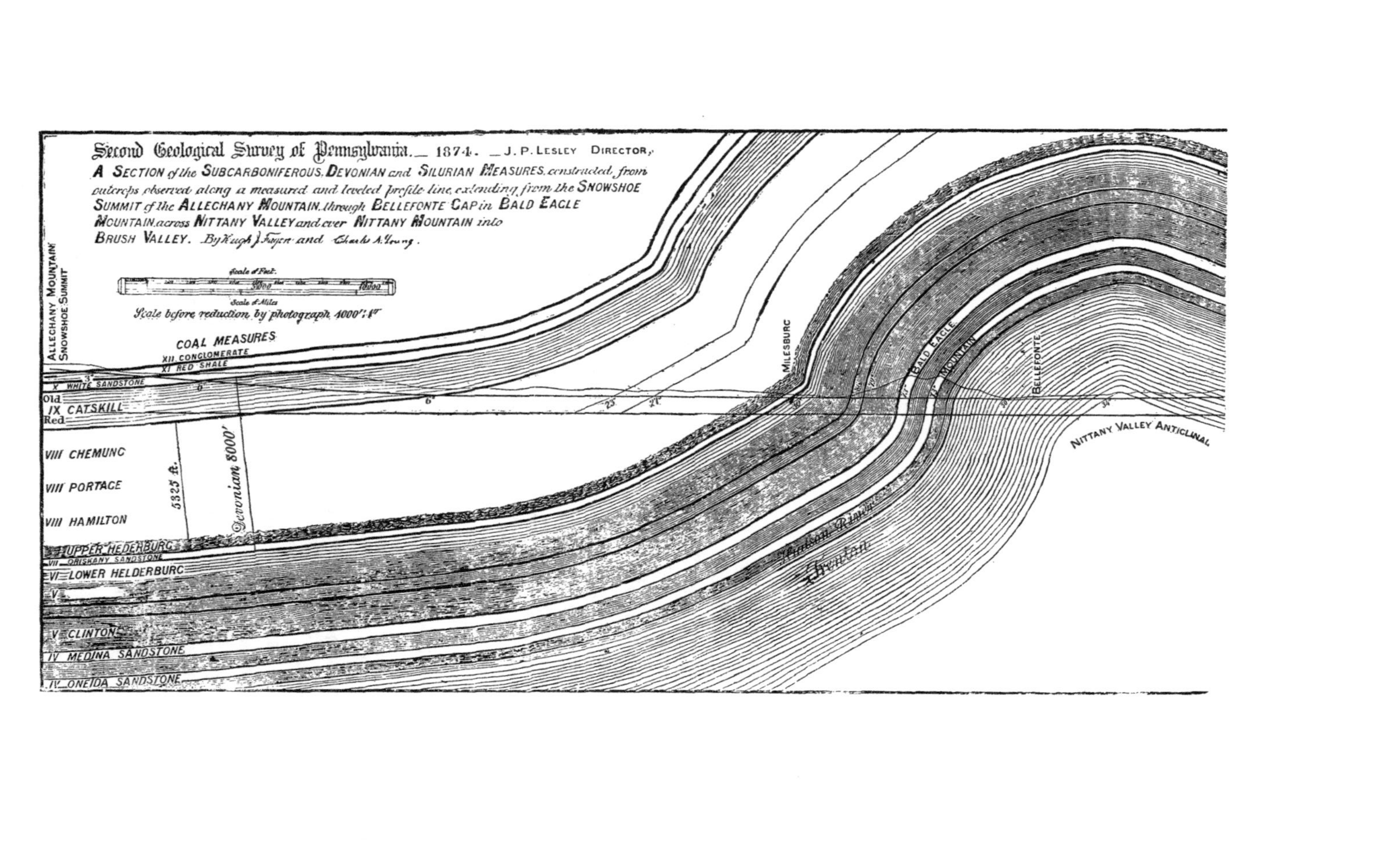
Second Geological Survey of Pennsylvania. — 1874. — J. P. Lesley Director,
A Section of the Subcarboniferous, Devonian and Silurian Measures, constructed from outcrops observed along a measured and leveled profile line, extending from the Snowshoe Summit of the Allegheny Mountain, through Bellefonte Gap in Bald Eagle Mountain, across Nittany Valley and over Nittany Mountain into Brush Valley.
Scale of Feet.
5000
10000
Scale of Miles
Scale before reduction by photograph 4000′:1″
Allechany Mountain Snowshoe Summit
Coal Measures
XII Conglomerate
XI Red Shale
X White Sandstone
Old
IX Catskill
Red
VIII Chemung
VIII Portage
VIII Hamilton
5325 ft.
Devonian 8000′
Upper Helderburg
VII Oriskany Sandstone
VI Lower Helderburg
V
V Clinton
IV Medina Sandstone
IV Oneida Sandstone
Milesburg
Bald Eagle
Mountain
Bellefonte
Hudson River
Trenton
Nittany Valley Anticlinal

REPORT OF PROGRESS

IN THE

CLEARFIELD AND JEFFERSON

BITUMINOUS COAL DISTRICT,

1874.

BY FRANKLIN PLATT, ASSISTANT GEOLOGIST.

CHAPTER I.

A Sketch of the Geography and Geology of the Bituminous Coal Fields of Pennsylvania in relation to the Clearfield and Jefferson District.

This subject has been treated in so masterly a manner by the State Geologist of the First Survey of Pennsylvania, Prof. H. D. Rogers, in the Second Volume of his Final Report of 1858, pages 795 to 811,* that it will only be necessary to state shortly what parts of Western and Northern Pennsylvania are occupied by Coal Measures,—the geographical relationship of the Clearfield and Jefferson District to the whole area so occupied—and the series of rocks constituting the Coal Measures, considered as one formation, the last and highest of the Palæozoic or Old-Age Geological System.

The Anthracite Coal Fields are confined to Eastern Pennsylvania.

The Semi-bituminous Coal Field of Broad Top in Bedford and Huntingdon counties is wholly isolated from the Western Bituminous Coal Field.

The Semi-bituminous Coal Basin of Cumberland, in Maryland, has its northern end only within the geographical limits of Pennsylvania, in the south-east corner of Somerset county; is also isolated from the main Coal Field like the Broad Top basin, lying nearly in range with it, but further south.

* See appendix A to this volume.

With these exceptions all the true Coal Measures of Pennsylvania are to be found behind the crest of the Allegheny Mountain; to the north of it as far as the New York State line; and to the west of it as far as the Ohio State line, and to the south-west as far as the Virginia State line.

But over this large area the coal rocks are by no means equally distributed. There has been great destruction of coal measures, in the lapse of ages, since that general uplift of the continent which put an end to the creation of coal beds, drained the old coal swamps, and raised the formation to a height of several thousand feet above the sea level. Ever since that event, through all the subsequent ages of New Red, Cretaceous, Tertiary and Recent, frost and water have been eroding the Coal Measures. The Susquehanna River Branches have carried cubic miles of coal from the surface of the northern counties into the Atlantic. The Tioga and Genesee rivers have done their share towards the removal; and the Allegheny and Monongahela rivers have furnished the Ohio river with still larger quantities of coal-waste for the delta deposits of the Mississippi.

As Middle Pennsylvania and Middle New York were lifted much higher above the old sea level, than South-west Pennsylvania, Ohio and Virginia were, the destruction of the Coal Measure Formation has been greater towards the north and north-east, and less (gradually less and less) towards the south-west.

Only the lowest bed of coal, or the two or three lowest beds, have been left, in isolated patches, on the tops of the mountains in Wyoming, Sullivan, Lycoming and Clinton, in Bradford, Tioga,* Potter, Cameron, M'Kean and Warren counties. Crawford and Erie have lost all their former coal.

The great productive bituminous coal field may, therefore, be properly said to commence (going south-west) in a belt of counties made up of Clearfield, Jefferson, Clarion, Venango and Mercer, stretching from the crest of the Allegheny mountains to the Ohio State line, a distance of about 140 miles.

All the counties to the south of this belt form an unbroken area of Coal Measures, passing over into Ohio and Virginia, and through Kentucky and Tennessee into Alabama.

*The Blossburg basin in Tioga county is exceptionally deep, and has preserved most of the lower productive coal measures.

The next belt includes Cambria, Indiana, Armstrong, Butler, Lawrence and Beaver.

From all the counties of these two belts, and from Cambria and Somerset,* eastern Westmoreland, and eastern Fayette, the Upper Productive coal measures only have been swept away. The *Lower* Productive coal measures are, in the main, left in these counties. There has been a good deal of even the lower measures carried off, but not so as to destroy the continuity of the coal field, except along certain narrow lines, marking the crests of anticlinal waves or ridges in the earth crust. Along these anticlinal waves the destruction or erosion has, of course, been greatest. Between them lie the synclinal " coal-basins," to be described in this report.

The counties which remain to be mentioned, lying south of the Conemaugh (Kiskiminitas) river, and the Ohio, have retained all their Lower Productive coal measures, and much of the Upper Productive coal measures. They are Allegheny, Washington, Greene, western Westmoreland, and western Fayette. Washington and Greene have suffered least, and have even kept a thick pile of strata over-topping even the Uppermost coal beds.

The geological map of the State accompanying Walling and Gray's Topographical Atlas of Pennsylvania† will make the above short statement sufficiently clear.

Productive Coal Measures.

The term " Productive Coal Measures," used above, has become a necessity from the fact that since the original geological classification was adopted from the English geologists, and applied with many needful modifications to American geology, the coal fields of the world have been extensively studied, and the views of geologists been somewhat altered. At first the " Coal Measures" were supposed to be of one age. They were divided simply into Upper and Lower. But in Nova Scotia, in Kentucky and Virginia, and elsewhere, *still lower* coal measures have been found. These *still lower* coal measures have not often been pro-

* Around Salisbury, along one small ridge, the upper productive coal measures have been spared. Another minute patch has been left in eastern Westmoreland, in the Ligonier Valley.

† Published by Stedman, Brown and Lyon, Philadelphia, 1872. The Survey will be prepared to publish a new geological map of the State when all the needful corrections are made.

ductive, neither in Europe nor in America, although they are so in eastern Kentucky, in West Virginia, and in Montgomery county in Virginia. But geologists must classify the formations according to the principles of the science, and not according to the commercial value of their minerals. Hence the terms Upper, Middle and Lower Coal Measures have come into vogue; and what once were the *Lower* Coal Measures of Pennsylvania, should now more properly be called the Middle Coal Measures. They still, however, continue to be the lower *productive* coal measures; for the workable beds of the *still lower* measures in Kentucky and Virginia are practically worthless in Pennsylvania.

Many years ago J. P. Lesley published a description of the two workable beds of coal in Montgomery county, Virginia, and his discovery of more than a dozen coal beds *under the red shale* along the western flank of Peak Mountain in Wythe county, next adjoining Montgomery, to the south of the New or Kenawha River. None of these beds were a foot thick, and most of them only two or three inches; but with their inclosing rocks they constituted a true "coal measures" several hundred feet in thickness, occupying the geological place of the Vespertine, White Catskill, No. X.

Within a few weeks of the writing of this (June, 1875,) Mr. Ashburner reported his discovery of *eleven* small coal beds distributed through the body of the White Sandstone of X. in the tunnel of the East Broad Top Railroad, in Sideling Hill, Huntingdon county, Pa., one of which is one foot thick. This bed he has also found exposed in a neighboring gap through the mountain. It will be seen in a subsequent chapter that this important discovery not only establishes the fact of a Lower Coal Measure proper in Pennsylvania; but explains the presence of two coal beds along the face of the Allegheny mountain, far beneath the conglomerate.

To make the affair more complicated there are coal measures far older and lying *lowest of all*, many thousand feet below the geological horizon of those of No. X, namely: near the bottom of No. VIII. These are well developed (geologically) only on the Juniata River, in Perry and Juniata counties, and although they are absolutely worthless for all mining purposes, they are nevertheless real Coal Measures; a formation constituting a fourth and oldest member of the series, near the bottom of the Devonian System.

To meet the difficulty Professor Rogers very wisely suggested the substitution of two geographical names for the epithets "Upper" and "Lower," as applied to the Productive Coal Measures, viz: The Monongahela River Series; and the Allegheny River Series. There is the best of reasons for these names because the *upper productive coal beds* crop out and are mined in Westmoreland, Greene and Washington counties, in the water basin of the Monongahela River; while the *lower productive coal beds* crop out along the Allegheny River, and occupy the hills which separate its numerous branches.

The Pittsburg Coal bed is the base of the Monongahela Series.

The Great Conglomerate is the base of the Allegheny Series.

All coal measures underlying the Conglomerate were at first called *False Coal Measures.* Of late they have been called *Sub-Conglomerate* coal measures. Some have wished to call them *Sub-Carboniferous* coal measures; but this is inaccurate, seeing that they are the lowest part of the Carboniferous System, and not at all underneath it. The Nova Scotia geologists would call these beds the *Lower* Coal Measures; the Allegheny beds the *Middle* Coal Measures; and the Monongahela beds the *Upper* Coal Measures.* In the Final Report of 1858 they are called the

---

* The Coal Measures of Nova Scotia are thus subdivided by Prof. J. W. Dawson, of Montreal.

"1. The Upper Coal Formation, containing coal formation plants, but not productive coals.

2. The Middle Coal formation, or coal formation proper, containing the productive coal beds.

"3. The Millstone Grit Series, represented in Nova Scotia by red and gray sandstone, shale and conglomerate, with a few fossil plants and thin coal seams, not productive.

"4. The Carboniferous Limestone, with the associated sandstones, marls, gypsum, etc., and holding marine fossils, recognized by all paleontologists who have examined them as Carboniferous.

"5. The Lower Coal Measures, holding some, but not all, of the fossils of the Middle Coal formation, and thin coals, not productive; but differing both in flora and fauna from the Upper Devonian, which they overlie unconformably.

"The limits of these groups are however in most cases not clearly defined."

These carboniferous rocks of Nova Scotia are given by Prof. Dawson as 16,000 feet thick; the Millstone Grit Formation alone being given as 5,000 to 6,000 feet thick.

Without attempting here to compare the different groups of the carboniferous rocks as exposed in Nova Scotia and in the Appalachian Coal Basin, it

Sharon Series; because one of the beds has been extensively worked at Sharon, in Mercer county, on the Shenango River. Mercer county is in fact the only county in the State* where any valuable coal bed has been found underneath the Great Conglomerate; but in Eastern Ohio the Sharon Coal is one of the most valuable beds.

If the geographical nomenclature, then, be insisted on, it might be well to call the sub-conglomerate coals the *Shenango River Series.* This would allow the No. X coals to be called the *New River Series;* and the older Devonian coal beds of No. VIII, to be called the *Juniata River Series.*

But as the numbering of the formations in the First Geological Survey of 1835–1841, is as convenient and useful now as ever, and as the Juniata Coal beds in Formation No. VIII are of no value, and are scarcely ever mentioned except in connection with the geological structure of that region; and as the Sharon coal falls in No. XI, and can still be as it has always been called the coal of XI, there seems to be no sufficient reason for abandoning the old division of our Pennsylvania Coal Measures into simply *Upper* and *Lower*, adding the term "Productive" to prevent the possibility of mistake.

The Lower (productive) Coal Measures, then, cover all Western Pennsylvania within the limits above described.

The Upper (productive) Coal Measures are confined to Allegheny, Westmoreland, Washington, Greene and Fayette; except that two long thin points are projected northwards into Armstrong and Indiana counties; one little patch remains in Ligonier

may be said that the proposed subdivision of the Carboniferous rocks into Lower, Middle and Upper Coal Measures is not only applicable to the facts as exhibited in the Appalachian Basin but also more in harmony with the grouping adopted by Prof. Dawson in Nova Scotia.

It may be worth noting here that an article in a recent number of the New York Engineering and Mining Journal refers to a report of Mr. D. P. Murphy on the Tecomatlan, Olamatlan, and Pena de Ayuquila coal basins in Mexico. Mr. Murphy considers the measures exposed to be true Carboniferous rocks, yielding *neuropteris gigantea, hypecopteris longetica,* &c. He subdivides the measures into Upper, Middle and Lower Coal Measures: the Upper Coal Measures containing the workable coals and the Middle Coal Measures containing the characteristic Millstone Grit or Conglomerate.

* Unless the coal beds at Shamokin and the new Lykens Valley Coal bed, *inside* the Conglomerate of the Anthracite Field, represent the Sharon Series; which has not yet been proved.

Valley, and a small but important basin of it has been left at Salisbury, in Somerset county.

The following report on Clearfield and Jefferson can have nothing to say about the Upper or Monongahela Series, for it has been entirely removed by the erosion of the Moshannon, West Branch, Bennett's Branch, Sandy, Red Bank, Mahoning, and other streams mentioned in the Report.

The Lower (Middle) or Allegheny Series of coals occupy the entire district, except along the anticlinals, where they also have been partially or entirely washed away, leaving the Great Conglomerate (No. XII) bare. Bennett's Branch has even worn through this formation, and its banks and bed reveal the red shales of XI, the white sandstone of X,* the red shales and sands of IX,† even as low as the top measures of VIII.‡

Scheme of Coal Measures.

On the following page a scheme of the Coal Measures and Underlying Formations is placed for reference.

The names given to the formations by the New York Geologists previous to 1843, are prefixed. The nomenclature invented by Professors H. D. & W. B. Rogers, and used throughout the Final Report of 1858, is affixed to the Numbers (from I to XII,) used by the Geological Corps of the First Survey, and in the Annual Reports of 1836 to 1842.

The letters A, B, C, D, E, prefixed to the five principal coal beds of the Allegheny Series, are those adopted by Hodge and Lesley, in studying the series through Lycoming, Clearfield, Cambria, Somerset and Fayette in 1840, and used in the Fifth Annual Report of 1841, in Lesley's Manual of Coal in 1856, and in Prof. Rogers' Annual Report of 1858. They are generally adopted by the people of Western Pennsylvania, and cannot now be changed, however desirable it has become to do so, inasmuch as several locally important intermediate coals are omitted from this series of lettered beds. These letters, it must be understood, only indicate the five *principal* deposits of coal in the Allegheny Series. There has always been a difficulty in applying them to the beds actually outcropping on the Allegheny River; and as early as

* Upper Catskill. † Lower Catskill. ‡ Chemung.

1842 it was suspected that A of the Allegheny Mountains was B on the Allegheny River.

THE COLUMN OF PALÆOZOIC FORMATIONS.

Series.			
	Upper Barren measures.		
Monongahela, - -	M. Brownsville (Washington) Coal bed.		
	L. Waynesburg Coal bed.		
	K. Sewickley Coal bed.		
	J. Redstone Coal bed.		
	I. Pittsburg Coal bed.		
Conemaugh, - - -	Middle Barren measures.		
	Mahoning Sandstone.		
Allegheny, - - -	E. Upper Freeport Coal bed.		
	D'. Middle Freeport Coal bed.		
	Freeport Limestone.		
	D. Lower Freeport Coal bed, (Reynoldsville.)		
	Freeport Sandstone.		
	C. Kittanning Coal bed.		
	B'. Ferriferous Coal bed.		
	Ferriferous Limestone		
	B. Clarion Coal bed.		
	A. Brookville Coal Bed.		
	Conglomerate, - - - -	No. XII.	(*Seral.*)
Shenango, - - - -	Sharon Coal beds, - -	No. XI.	(*Umbral.*)
	Red Shale, - - - - -		
Catskill, - - - - -	New River Coal beds.	No. X.	(*Vespertine.*)
	White Sandstone, - -		
	Red Sandstone, - - -	No. IX.	(*Ponent.*)
Chemung, - - - -	Olive Shales, - - - -	No. VIII.	(*Vergent,*)
Portage, - - - - -	Olive Sandstones, - -		
Hamilton, - - - -	Juniata Coal beds, - -		(*Cadent.*)
	Black Shales, - - - -		
Upper Helderberg, -	Corniferous Limestone,		(*Postmeridial.*)
Oriskany, - - - -	White Sandstone, -	No. VII.	(*Meridial.*)
Lower Helderberg, -	Lewistown Limestone,	No. VI.	(*Premeridial.*)
Waterlime, - - - -	Cement Layers, - - -		(*Scalent,*)
Clinton, - - - - -	Red Shales and Fossil ore,	No. V.	(*Surgent.*)
Medina, - - - - -	Red Sandstones, - - -	No. IV.	(*Levant.*)
Oneida, - - - - -	White Sandstones, - -		
Hudson River, - -	Slates, - - - - - - -	No. III.	(*Matinal,*)
Trenton, - - - - -	Limestones, - - - -	No. II.	
Calciferous, - - -	Dolomites, - - - - -	No. II.	(*Auroral.*
Potsdam, - - - -	Sandstone, - - - - -	No. I.	(*Primal.*)

The Tionesta Series.

Still greater difficulties arose when the lettering of the Allegheny Mountain beds and of those in Somerset county was applied to the series of beds in Elk and M'Kean counties. It was in these Northern counties that the great mistake of the First Survey seems to have been made. Hodge in 1839, named and described the Sharon Series as *underneath* the Conglomerate, from his explorations west of the Allegheny river. Lesley in 1841, named and described the Tionesta Series as *overlying* the Conglomerate, as he approached the Allegheny river from Potter county. No one assistant of the First Survey studied both the regions; no one, therefore, was in a condition to discover what seems, by the surveys of 1874 in Venango county, to be the true state of the case, viz: *that the Tionesta Sandrock is the Conglomerate, and therefore the Tionesta Series of Coal beds is the same with the Sharon Series.*

The mistake arose from the fact that, in the unbroken wilderness east of the Allegheny River, and in the hasty reconnoisance survey of 1841, it was impossible to keep hold of the Conglomerate after it had begun to turn into a sandstone. Another conglomerate lies 200 feet below it in the M'Kean and Venango country, and is the principal rock of the country. This, which is Mr. Carll's Second Mountain Sand, was mistaken for the Conglomerate No. XII; and of course, the Tionesta Sandrock, (the real No. XII,) was supposed to be the first great Sandstone of the Allegheny Coal Measures.

With the light now possessed, it seems, therefore, necessary to expunge the whole Tionesta group, sandrock and coal beds, from Mr. Rogers' column. Certainly no such series exists between the Allegheny River and the Ohio State line; no such series can be recognized in the Southern counties; and no allusion to any such system will be found in this Report on the Clearfield and Jefferson district.

CHAPTER II.

The Subdivision of the District into Five Coal Basins, by anticlinal waves.

The First Geological Survey sufficed to show that the Bituminous Coal Measures lie in six principal basins, arranged in concentric curves of wide sweep, entering the State from Eastern and Middle New York and leaving it across the Maryland and Virginia State line.

The five anticlinal waves separating these six basins were thought to be continuous the whole distance, and were so laid down upon the original Geological map of the State. But the Survey of 1874 has shown that one at least consists of separate anticlinals, the ends of which lap past each other. The general persistency of these grand anticlinals however is very remarkable, as well as their general smoothness and regularity of curvature. They were first numbered by Mr. J. D. Hodge, who called the basin just back of the crest of the Allegheny Mountain the *first basin,* and the anticlinal next west of it the *first axis.* The Second basin and Second axis, third basin and third axis, &c., follow in regular order, going west, or north-west; the fifth axis crossing the Allegheny River below the mouth of Red Bank, and the Ohio River three miles below Pittsburg. All west of this axis has been called the sixth basin. But smaller axes have been discovered in the Beaver river country, and others in Ohio, which subdivide the sixth basin into several.

In this report of Clearfield and Jefferson county a belt of country crossing five of these basins will be described, viz:

The First, or Philipsburg—Snow Shoe Basin.

The Second, or Clearfield Basin.

The Third, or Luthersburg—Bennett's Branch Basin.

The Fourth, or Reynoldsville—Punxatawny Basin.

The Fifth, or Brookville—New Bethlehem Basin.

CHAPTER III.

The First Bituminous Coal Basin and First Anticlinal Axis.

The Anticlinal Mountain of Laurel Hill along the broad flat summit of which runs the nearly straight county line between Somerset on the east, and Fayette and Westmoreland on the west, has hitherto been taken as the western boundary of the First Bituminous Coal Basin.

But in Somerset county, Negro Mountain, coming up across the Maryland Line, sub-divides the First basin into the Turkey Foot or Ursina-Johnstown sub-basin, and the Salisbury-Berlin sub-basin. On and south of Castleman's River these are entirely distinct basins, the coal beds not passing over the anticlinal from one into the other. But north of that river, around Somerset, and as far north as Stony, Shade and Paint Creeks, the distinction disappears and the coal beds spread, nearly flat, across the anticlinal, over the country. The Berlin sub-basin can, however, be indistinctly traced northward until it vanishes along the crest of the Allegheny Mountain at the heads of the South Fork of the Conemaugh.

Another and independent low anticlinal crosses the Conemaugh six miles above Johnstown, at the Viaduct and Big Bend, which again sub-divides the First Basin into two; but only geologically, for the coal beds sweep over this anticlinal. The Viaduct anticlinal has never been carefully traced southward; but it seems to die out in the centre of the Ursina sub-basin. Northward, in Cambria county, it is wholly unknown as yet; but the Ebensburg country is being carefully surveyed (July, 1875,) for the purpose of tracing it into Clearfield county.

It is not known, therefore, nor will this report assert, that the First sub-axis, west of Philipsburg, is a continuation northward of the Viaduct axis: although so far as the investigation has been carried it has become more and more apparent that this is probably the case. The work of 1875 in Cambria county will definitely settle this question.

The Laurel Hill axis in Clearfield county is shown on the map accompanying this report, (Plate IV.) It crosses the Conemaugh just west of Johnstown, the Black Lick Creek west of Ebensburg, Clearfield Creek several miles above its mouth, and the West Branch of the Susquehanna west of Frenchville, about 5 miles west of Karthaus. After crossing the Black Lick Creek it loses gradually its character as a straight, well defined ridge: and where last seen in Clearfield county, crossing the West Branch of the Susquehanna River, is much flattened out and seems to be dying gently away to the northward.

The First sub-axis in Clearfield county was found to be certainly not a straight line. One of the maps accompanying this report, (Plate IV,) will show that it is made up of three lines, the intermediate one lying somewhat to the west of the two end lines.

It is made out, then, by the work of 1874 that we are no longer to look upon the First sub-axis as one long continuous line of upheaval; but that it consists of numerous anticlinals arranged *nearly* in one continuous curve, but not quite.

The run of the First Axis through the yet unbroken wilderness of Beach Creek is still unstudied. All we know is that the First Basin, on the Susquehanna West Branch, contains the Tangascootac and Queen's Run Coals; the First Axis being well marked in the river bluffs 8 miles above Farrandsville. It can be traced across Pine Creek, to Ralston on the Lycoming, and comes out in the broad valley of VIII (Chemung) below Wyalusing Falls on the North Branch Susquehanna. J. T. Hodge traced it all this distance in 1840. Thence it passes on through Montrose and Great Bend, and has been traced by Prof. Hall and Mr. Sherwood to the Schoharie Creek in Eastern New York.

All the coal south and east of this axis was placed by Mr. Hodge in the First Bituminous Coal Basin, and is so described in the Final Report of 1858. But we have seen that in Somerset county there are two basins; in Cambria county two sub-basins, and in Clearfield county two sub-basins. It is quite possible, therefore, that Hodge's *First Basin* in Lycoming and Clinton corresponds to the Phillipsburg-Osceola basin alone, and that his *Second Basin*, to the north-east of the district under report, corresponds to the *second sub-division* of the *First Basin* in Clearfield and Cambria, to the south of it.

CHAPTER IV.

The Second Bituminous Coal Basin and Second Anticlinal Axis.

The south-east boundary of the Second Basin is the First Axis, as described in the last chapter.

The north-west boundary of the Second Basin is the Second Axis, which elevates Chestnut Ridge in Westmoreland and Fayette counties.

Between Laurel Hill and Chestnut Ridge lies the long strait valley of Ligonier south of the Conemaugh, and of Armagh north of the Conemaugh.

The First Axis can be best studied in the deep gaps of Laurel Hill below Confluence (Ursina, Turkey-foot) in Somerset, and below Johnstown in Cambria. It is plainly marked also in the shallower gorges made by the South Branch of Black Creek west of Ebensburg; by the North Branch of Black Lick north-west of Ebensburg; by the Susquehanna River in Carroll township, Cambria county; and by Chest Creek near the north line of Cambria county.

The Second Axis can be studied in the Youghiogheny Gap near Connellsville, the Loyalhanna Gap near Youngstown, and the

Conemaugh Gap near Blairsville. All three gaps cut the mountain down to its base, a depth beneath the crest of from 1,000 to 1,300 feet. The arch is symmetrical, and all the formations from XII to VIII are exhibited. The axis is again finely shown in the almost equally deep gap of Black Lick Creek (5 miles north of the Blairsville Gap); in the gorge of Yellow Creek, (15 miles) in the gorges of Penn Run (18 miles) and Two Lick Creek (20 miles.) For the rest of its course in Indiana county, about 20 miles, it has not been studied, neither during the First Survey of the State, nor in 1874; but it certainly runs on to the north-east corner of Indiana county, and enters Clearfield county in Bell township.

On Map, plate VI, bound with this report, the approximate crest line or axis of this Second Anticlinal is drawn through Bell township, through the north-west corner of Penn, and south-east corner of Brady, across Little Anderson creek and Anderson creek one mile above their junction, and so on across the north line of Lawrence township; where it leaves Clearfield county and enters Elk.

Before reaching Anderson creek it runs a little west of Packersville or Forest P. O., and after passing the creek it runs a mile or more east of Johnson's mills, or Rockton P. O., as will be described more in detail in chapter XIX of this report.

The Second Axis elevates Chemung rocks (No. VIII) in the bed of the Sinnemahoning at the mouths of its three principal branches, Bennett's, Driftwood, and East Branch. It then traverses the wilderness upland of south Potter county, and issues thence east of Pine creek to make the broad valley of Wellsborough and Mansfield in Tioga county, *north of the Blossburg Coal Basin.* Then it spreads out the Chemung Formation (No. VIII) over all northern Bradford; crosses the North Branch of the Susquehanna at Athens, and keeps on through the southern counties of New York towards Albany.

Now this is Hodge's Third Axis.

Where then shall we look for his Second Axis? We find it at Towanda, 15 miles further south than Athens. Following this backwards, north of the Towanda coal field, towards the west and south-west, we see it running *south of the Blossburg Coal Basin*, and north of the Ralston coal field; crossing Pine creek half way between the First and Second Forks, and the Susquehanna West Branch at Hyner's and Ritchie's Stations on the P. and E. R. R. below Youngwomanstown. For the next 24 miles (going south-west) it elevates the unexplored wilderness mountain-land from which descend all the waters of Beech creek. It crosses the pike between Snowshoe and Karthaus, and may be seen drawn in the north-east corner of Map Plate VI, as a black line, one mile east of Kylertown in Clearfield county; where it ends, or is set back north-west several miles, and then runs on to Amesville; here it is again set forward to its old place near Janesville; and so keeps on south through Cambria county to the Viaduct on the Pennsylvania railroad, as described in the last chapter.

It is evident then, Hodge's Second Axis in the north-east is far from being the Second Axis of the southern counties. It seems to be the *First sub-axis* which splits the First Basin in all. the country south-west of the Susquehanna River.

The First Axis proper has been described in the last chapter, as dying down to the west of Karthaus. From our present knowledge, we may even consider it as forming one with the Second Axis, where they cross the Sinnemahoning near the mouth of the East Branch. Until careful *instrumental* surveys are made along the valley of the Susquehanna and Sinnemahoning, from Lock Haven to Ridgway, so as to fix the exact number, place and size of the anticlinal rolls which cross it, and similarly careful surveys are carried across the country to and along Pine Creek, it will be impossible to correllate the axis of the Northern with those of the Southern counties. But it seems as if the exceptional breadth of the Wellsboro' Valley in Tioga county, and the extraordinary outspread of the coal measures in southern Potter and northern Clinton, can be explained by the alliance of the First and Second anticlinals in Clearfield.

Between the First and Second Anticlinals lies the Second Bituminous Coal Basin of the Ligonier and Armagh valley, in Fayette, Westmoreland and Indiana counties; and the prolongation of this into Clearfield county makes the Curwinsville and Clearfield Basin. to be described in Chapter XVII below.

From the crest of Laurel Hill to the crest of Chestnut Ridge across the Ligonier Valley is everywhere almost exactly 10 miles.

From the First Axis on Clearfield and Little Clearfield creeks to the Second Axis on Anderson and Little Anderson creeks, across through Curwinsville is 15 miles. The Second Basin therefore broadens going north; but it also shallows somewhat.

A small sub-axis splits the Ligonier basin into two. It passes near Bolivar, and close to the foot of Chestnut Ridge, sharpening the dips (east) in the ridge at one place to eighty (80° !)

Whether such subordinate rolls subdivide the Second Basin in its run through Indiana and Clearfield counties is not yet discovered.

CHAPTER V.

The Third Bituminous Coal Basin and Third Anticlinal Axis.

The Second Axis is everywhere the south-east border of the Third Basin, which consequently lies on the west flank of Chestnut Ridge in Fayette and Westmoreland counties, where it is so deep as to take in the Pittsburg coal bed and several higher beds of the Monongahela Series; and in Indiana county, where it is deep enough to hold most of the Barren Measures, but not up to the Pittsburg coal. In Jefferson and Clearfield it gets to be so shallow that it only retains the Clarion Series, and does not even have the Lower Freeport Coal Bed. It is everywhere a very narrow basin, varying from five miles at Blairsville, to seven at Luthersburg. It holds so *much* of the Coal Measures in the country south of the Conemaugh, because its dips are steep. It holds so *little* in Clearfield county, beçause its dips are very gentle, hardly removed from the horizontal.

The Third Axis is its north-west limit. This axis crosses the Youghiogheny below Connellsville, the Loyalhanna just above New Alexandria, the Conemaugh (Kiskiminitas) two miles below Blairsville; runs nearly under the county town of Indiana; enters Jefferson county three miles south-east of Punxatawney; and Clearfield county (See Plate VI, H,) at Evergreen Station, on the Low Grade Railroad and Sandy Lick Creek.

The perfect straightness and regularity of this remarkable anticlinal is as wonderful as any phenomenon in geology. From where it enters the State across the Virginia line, to where it leaves Clearfield and enters Elk county, it runs *one hundred miles in an absolutely straight line.*

Here it enters and runs through the length of Boon's or Elk Mountain; crosses successively Trout run and Hick's run to Emporium on the Sinnemahoning; penetrates the wilderness south of Coudersport; and after passing Pine creek at the mouth of the Genesee Fork, enters Tioga county to form the Chemung (No. VIII) valley of the Cowanesque; crosses the Tioga river below Lawrenceville, and runs on past Elmira, in New York State, towards Albany. This Third Axis of the southern counties, and of Clearfield, seems to be the Fourth Axis of the north. The only doubtful region of its identification is from Elk to Potter counties; for the Old Survey turned this axis short before reaching Emporium, and made it run straight to Wellsboro', in Tioga county.

The straightness of the Third Axis in Elk, M'Kean, Potter and Tioga counties, a distance of seventy-five miles, is nearly as remarkable as its direct allignment from Elk county to Virginia. What we are to think of the force and mode of action productive of an anticlinal rock wave, continuous and almost rectilinear for about two hundred miles, it is difficult to say.

The Third Bituminous Coal Basin holds the towns of Unionville, Connellsville, Latrobe and Youngstown, Blairsville, Homer, Kintersburg, Robertsville, Washington (in the north-east corner of Indiana county,) Troutville and Luthersburg in Clearfield county, and the Bennett's Branch valley, with the stations along the Low Grade Extension of the Allegheny Valley railroad to Driftwood. It may therefore be called indifferently the Connellsville, the Blairsville, the Luthersburg, or the Bennett's Branch Basin. Its coals around Luthersburg are described in Chapter XIX; and its northern extension, down Bennett's Branch and the Railroad, in Chapter XXI.

CHAPTER VI.

The Fourth Bituminous Coal Basin and Fourth Axis.

North-west of the Third Axis last described, lies the Fourth Basin, of Reynoldsville and Punxatawney in Jefferson county (described in chapters XXII, XXIII and XXIV;) of New Alexandria and Greensburg, in Westmoreland county; and of New Salem, in Fayette countv. Also the coal field of Coudersport, in Potter county.

The Fourth Axis, bounding this basin on the north-west, is rather obscure on the Monongahela; but very plain between Greensburg and Adamsburg (Westmoreland), and thence northward, on the Kiskiminitas above Saltzburg. It has not been carefully traced through the Western townships of Indiana county, but is plainly marked on the Mahoning between Nicholsburg and Perryville, and on the Sandy Lick at Port Barnett, Jefferson county, (see Map, Plate VII, of this report;) whence it runs to the P. and E. R. R., between Ridgway and St. Mary's; and to the head waters of the Genesee river and the north-east corner of Potter county. It is traceable far into Middle New York State. Much of the country through which this axis runs has yet to be studied instrumentally before it can be well understood. Elk and M'Kean counties require and deserve a long, careful and laborious survey.

But a more important investigation is required in Indiana and Armstrong counties, to settle the question whether the Saltzburg axis be really that at Nicholsburg. It is possible that the prolongation of the Nicholsburg Fourth Axis is that which crosses the Kiskiminitas seven miles below Saltzburg, above Warrentown, and keeps on across the Youghiogheny and Monongahela river at Monongahela City, and has beeen traced recently (July, 1875,) by Prof. Stevenson in a straight line through Greene county, 4 miles east of Waynesburg. If so, *the Saltzburg axis is a sub-axis splitting the Fourth basin.*

What renders this probable is the great breadth of the Fourth basin in Jefferson county, from Evergreen across to Port Barnett, 12½ miles. The Fourth Basin is sub-divided by a well marked anticlinal sub-axis just west of Reynoldsville.

Nothing but a long continued accurate instrumental survey of the neglected region lying north of the Kiskiminitas and east of the Allegheny, as far as Indiana and Punxatawney, will suffice to clear up this geological question lying at the bottom of the future material interests of that section.

CHAPTER VII.

The Fifth Bituminous Coal Basin and Fifth Axis.

North-west of the Fourth anticlinal lies the Fifth Basin of Brookville and New Bethlehem in Jefferson and Clarion counties (described in Chapter XXV;) of Kittanning, Freeport and Leechburg in Armstrong county; and of the Pittsburg, M'Keesport and Washington country. In it lie also the detached coal fields north and west of the Clarion, around Ridgway and towards Smethport.

The Fifth anticlinal axis, bounding it on the north-west, comes up from the south-west, out of Washington county, in a straight line; crossing the Ohio 3 miles below Pittsburg; the Allegheny half way between the mouths of Red Bank and Mahoning; the Red Bank at Lawsonham Mills; and the Clarion below Jefferson. It no doubt continues through M'Kean county west of Smethport, and into New York State.

This is an even more remarkable straight and continuous run than that of the Third Axis, for it must measure nearly 150 miles, without any such bend as at one point characterizes the other.

The Fifth Basin contains the Brookville coals in Jefferson county, and those mined along the Red Bank Valley in Clarion county, described in Chapter XXV of this report. Nothing need here be said about the Sixth Basin, no part of which comes into this district of Clearfield and Jefferson county, since it embraces all the rest of the Bituminous Coal Field as far as the Ohio State Line.

CHAPTER VIII.

Vertical Sections in the text.

In the following chapters the explorations of 1874 in the several basins, beginning with the First and ending with the Fifth, will be described in detail, and in as regular an order as the general outspread of the measures over all the basins and their separating anticlinals will permit.

Numerous vertical sections were made, not only of the measures exposed on the hill sides, but of the coal beds with their partings and roof rocks, where mining operations have been or are now going on. These sections are not only given in the text *always in natural descending order*, that is from above downwards, but in a picturesque form, as wood cuts indented in the page, for the convenience of the reader, and to make the detection of errors easy.

No elaborate attempt is made to collate and compare these sections one with another, to harmonize them, to obtain a complete system of coal beds, still less to work out the geological questions suggested by their frequent variations from one another. This can be better done after another season's work, doubling the number of facts at hand, and the breadth of the field under survey.

Some few points, such as the marked absence of limestone beds from, and the equally marked abundance of coarse sandrocks in, the coal measures of the First Basin, will be apparent at a glance.

Some inconsistencies in the sections, when compared, must be attributed to defective leveling, sometimes with an ordinary barometer, and at other times with the hand level, spirit level and vertical circle. But allowance being made for such errors, there remains sufficiently striking evidence of the thickening and thinning of the coal beds, and of the parting rocks, in very short distances.

In the First Basin, for example, while a comparison of the Snow Shoe Section, Fig. 31, at the furthest north-east point examined, with the Section, Fig. 4, on the Upper Moshannon, the furthest point to the south examined, shows a remarkable similarity of measures, there occur variations, some slight and some very decided, in the interval between these two points, distant apart about 35 miles.

In the Fourth and Fifth Coal Basins there is very considerable increase in the actual thickness of coal measures, as the sections plainly show

The nomenclature adopted in the Final Report of Rogers, for

the most important rocks and coal beds is now widely known and entirely adopted in Pennsylvania; and it would seem decidedly inadvisable to make any changes except such as may be forced on us as absolutely needed. The coal beds in these sections lying between the Millstone Grit and the Mahoning Sandstone, are lettered A, B, C, D and E;—D and E representing the Lower and Upper Freeport Beds. In some of the sections the precise identification is left for future examination and revision; and it may be that some changes will eventually be needed. In every case where a bed can possibly be identified under one of the present well known names, care will be taken to avoid giving it any local name.

In the section at Snow Shoe in the First Basin, (Fig. 31,) Bed A is the first workable bed above the Millstone Grit or Seral Conglomerate. Beds D and E are the Freeport Beds. This section agrees well with the Karthaus section (Fig. 37) in the second sub-division of the First Basin, re-produced from the Final Report. The only change made is, that while in the Final Report the small rider of coal on top of the Freeport Limestone is called the Upper Freeport Bed, in the revised section (Fig. 37,) as well as in the Snow Shoe section (Fig. 31,) the valuable coal bed 50 feet above the Limestone is called the Upper Freeport Coal Bed. The other sections made in the Second Basin also agree well.

CHAPTER IX.

Geology of the First Bituminous, or Steam Coal Basin of Clearfield County.

The massive conglomerate at the base of the Lower Productive Coal Measures, the Seral Conglomerate of Rogers, crowns the straight and even crest of the Allegheny Mountain, and from that point dips gently down north-west into and under the First Bituminous Coal Basin. The coals, sandstones, fire-clays and shales of these Lower Coal Measures come in gradually on top of it.

The sketch map of part of the First Basin (Plate V,) and the cross section show the main features of the geology. The railroad line upon the map, and the positions of the collieries lying directly upon it, were kindly furnished by the chief engineer of the Tyrone and Clearfield Railroad. The Houtzdale and Mapleton Branch Roads, and the position of the collieries opened on or near them, were plotted from notes furnished by Mr. L. J. Lingle, of Osceola.

The lowest workable coal beds of the Lower Productive Coal Measures have their south-east outcrop just west of Sandy Ridge, but are not opened there. They are seen outcropping on the Tyrone and Clearfield Railroad between Sandy Ridge and Osceola, and are opened; and the second workable bed (B) is extensively mined, at Powelton. Though the coal beds seem to dip 3° or 4° at their exposures along the railroad, yet the entire amount of sinking to the north-west from Powelton, (1,784 ft. above tide,) to the outcrop above the railroad level at Osceola, (1,500 ft. above tide,) is not quite 300 feet. Just back of Osceola, on the hill top, and also on one extreme hill top south-east of the Moshannon creek, the upper bed (D) comes in, dipping slightly to the north-west. From the centre of the basin, which is here just north-west of Osceola, the measures rise to the north-west slowly but steadily, and for the most part come to daylight on the north-west side of the ridge on which lies the Goss Farm, three miles north-west of Osceola.

West of this Goss Ridge about 1½ miles is a higher wooded ridge which marks the axis of the First Anticlinal Sub-axis west of the Allegheny Mountain. South-westward along this ridge, the massive conglomerate rocks at the base of the Lower Coal Measures form the ridge crest for many miles; and north-eastward, these same rocks, breaking into huge masses, show plainly where they cross the Tyrone and Clearfield Railroad, a short distance north-west of Blue Ball Station. On this axis, as on the crest of the Allegheny Mountain and at other points noted during the season's work, the character of the Seral Conglomerate changes very decidedly and within a short distance; in some cases showing as a conglomerate with rounded pebbles of white quartz an inch or more in diameter, and within a few miles changed into a fine grained massive sandstone, not even conglomeritic.

The map shows how the waters of the Moshannon Creek collect

at Osceola, near the centre of the basin, and at Steiner's Station (near Philipsburg) at just about the centre of the basin. From each side of this centre line the coals rise in opposite directions; to the south-east to the crest of the Allegheny Mountain, and to the north-west to the high ridge of the First anticlinal sub-axis.

But while this is, broadly speaking, the average shape of the basin, it must be remembered that at scarcely any single point can the *actual average* slope of the beds be found for any considerable distance. As is always the case with nearly horizontal beds, there are gentle undulations, amounting in many cases to perhaps only a few feet, which throw the dip softly one way in one place and the other way in another; while the steady and main dip is always towards the centre or synclinal axis of the basin.

It should be noted that the First Basin at this point is decidedly *sinking to the north-east:* the rule in basins further west being a very slight sinking to the south. The result of this decided north-east sinking is that the basin here instead of being just parallel to the crest line of the Allegheny Mountain has for the centre line of its synclinal axis a general course of between north 55° and 60° east: the Allegheny Mountain at this point being about north 45° east to north 50° east.

This sinking to the north-east carries down the upper worked bed (D) so much that, while it is 152 feet above the Moshannon Creek at Osceola, it is only 50 feet above the water opposite Philipsburg. But the sinking stops abruptly just beyond that point, and the basin *rises to the north-east,* the lower rocks finally coming into the hill tops and confining the Lower Productive Coal Measures to a narrow basin between Philipsburg and the Snow Shoe. At Philipsburg the south-east outcrop of the coal is only 1½ miles west of the crest of the Allegheny Mountain. The rising to the south-west is very slight south of Osceola and ceases to show a short distance south-west of that place.

The vertical section, Fig. 1, was made at the deepest part of the basin and shows the measures exposed in the basin overlying the Conglomerate No. XII, the base of the Productive Coal Measures. It is partly compiled.

Before describing the measures exposed in this section, it

Fig. 1.
First Coal Basin

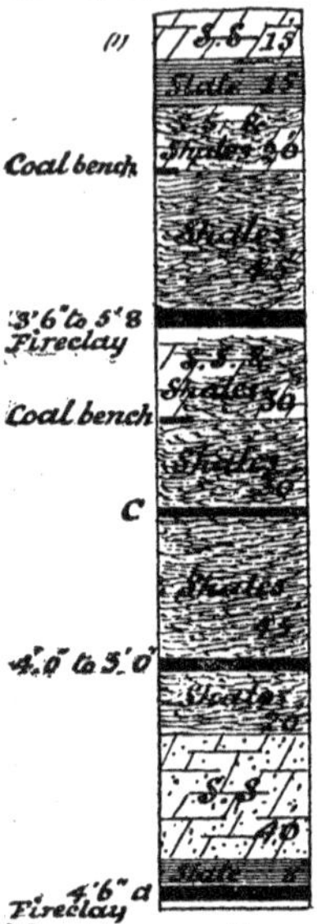

is proper to describe briefly the Conglomerate, No. XII, (Seral of Rogers,) lying at its base.

This well characterized and widely distributed member of the Coal Series consists, in its ordinary or typical condition, of massive beds of grey and whitish quartzose conglomerate alternating with grey and yellowish sandstone. In many districts it includes one or more seams of coal. It is an extremely persistent deposit and seems to underlie the Productive Coal Measures everywhere throughout Pennsylvania, except in Mercer and Crawford counties. Moreover it encloses not only beds of coal, identical in composition with the seams of the overlying rocks, but also innumerable fossil casts of genuine coal plants; especially the fragmentary stems of Lepidodendron, Sigillaria, and Calamites. The rock is fully described in the Final Report of Pennsylvania (Rogers) with its modifications of composition and thickness; from its exposures in the Anthracite Coal Regions of masses of coarse conglomerate in some places 1,000 feet thick, filled with quartz pebbles of large size, down to its appearance in the extreme west and north parts of the State as a siliceous white pebbly sandstone, and thinned down to 10 or 15 feet of total thickness. But it is not necessary to go a great distance in any direction to observe striking changes in its composition. Where the Fourth Anticlinal axis crosses Toby's creek in the north-east part of Jefferson county, this Seral Conglomerate makes the country rock, and covers the whole valley with massive boulders thickly studded with white quartz pebbles of the size of a hen's egg; while 12 miles to the south-west, directly along the same axis where it crosses the Sandy Lick Creek, scarcely a piece of conglomerate rock can be found. The rock is still a very massive sandstone, but without any conglomerate character.

No accurate measurements could be made in this Clearfield Steam Coal Basin of the total thickness of the Conglomerate; but judging from some imperfect measurements on the crest of the Allegheny Mountain at the Snow Shoe, and from the exposure at Karthaus in the western part of the First Coal Basin, the thickness can be put at from 200 to 250 feet.

Resting close on top of the upper layers of this Conglomerate, are the Sandy Ridge fire-clays; and over them sandstone, somewhat conglomeritic, not perfectly measured, but probably not less than 40 feet thick. Over this lies Coal Bed A, 4 to 4½ feet thick, a good strong steam coal, but in this part of the basin usually very sulphurous. This coal is at water level at Osceola on the Moshannon creek, and forms the base of the observed vertical section.

Over Coal Bed A, occur 20 feet of "brown rock," a rusty mass of thin sandstones, with an occasional massive layer, sometimes running into rusty shales and slates, in places with ore nodules, hematitic. The rock is noticeable and persistent through this part of the basin. Overlying this rock and extending up to the floor of Coal Bed B, is a hard massive sandstone, in some places conglomerate, but always fine grained, and much current bedded. This rock is in the main persistent in character, but not in thickness; thinning down very much in a distance of only five miles, and in one place being almost entirely replaced by thin sandstones and shales. Above Coal Bed B, the parting rocks up to the top of the section vary very much at different points, usually being made up of shales and thin slates, with a tolerably persistent show of sandstone underlying Coal Bed D, and another overlying Coal Bed D′. These variations will be found in the sections accompanying the detailed description of the basin.

The absence of limestone* from this First Basin, and the great prevalence of sandstone and very sandy shales have already been noted.

* For the purpose of comparison with sections to be given in the Second, Third, Fourth and Fifth Basins, the following tables are reproduced from the Final Report of the First Geological Survey, showing the conclusions reached from the mass of facts gathered by that Survey. The Cumberland Basin, for example, shows not more than ten feet of limestone; the First Basin in Clearfield has not more than seven feet at the utmost.

TABLE 1.

Showing the gradual increase in the aggregate thickness of the Limestones, as we cross the Southern Coal Field of Pennsylvania, westward.

Broad Top Basin, - - - - - - - - -	None.
Cumberland, (Potomac Basin,) about - - - - -	10 feet.

The decided columnar structure of the coals may also be noted as a marked feature of this First Basin.

Some of the sudden and striking changes in the measures and in the thickness of the coal beds are illustrated in the following vertical sections:

Fig. 2.

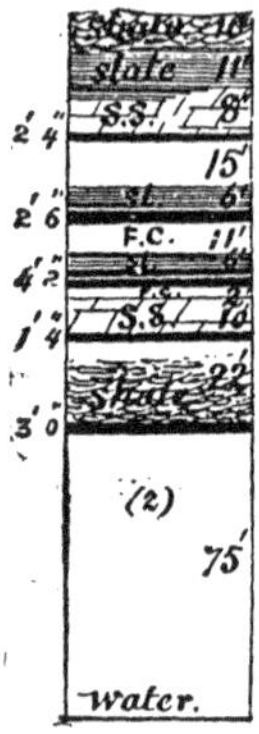

Fig. 2 represents a section made on the south-east side of the Moshannon Creek about a mile below Osceola and near Dunbar station. Identifying the lowest exposed coal as Bed C, the decided changes in Bed D and the measures above and below it, will be seen at once. The coals are only opened on the outcrop, but do not appear to keep up to the average in quality.

Fig. 3.

At Osceola the following section (Fig. 3) is exposed. The main features of the section have already been described in the section representing the average of the First Basin (Fig. 1.) Coal Beds A and B have been opened and worked on the south-east side of the Moshannon, but were both found decidedly sulphurous. Bed D is opened on the west side of the creek, back of the village and is an excellent 3½ to 4 foot coal.

Following up the Moshannon creek for 4 miles the following section (Fig. 4) is exposed. The variations in the section from the average

Eastern part of Somerset county, west of the escarpment, and about 12 miles west of the Cumberland Basin, - - - -	12 feet.
Ligonier Basin, 20 miles west of the last, - - - - -	30 feet.
Second Western Basin on the Youghiogheny, 15 miles west of the last, about - - - - - - - - -	40 feet.
Great Basin of the Monongahela and Ohio Rivers, - - -	60 feet.
Same Basin at Wheeling, about - - - - - -	200 feet.

TABLE 2.

Showing the Gradation in the thickness of the large Limestone Stratum, overlying the Pittsburg Coal seam.

Cumberland Basin, not more than - - - - - -	2 feet.
Eastern Somerset Basin, - - - - - - -	Thin.
Ligonier Basin, average about - - - - - -	7 feet.
Monongahela and Ohio Basin, - - - - - -	41 feet.
Same Basin at Wheeling, - - - - - - -	54 feet.

Fig. 4. *Upper Moshannon*

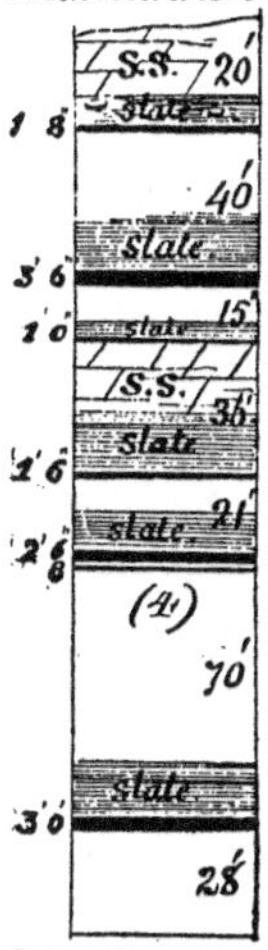

will be at once noted. Though the coal beds are only opened up by trial pits on the outcrops, yet from that imperfect showing there is reason to believe that in quality the coal of the Upper Moshannon region will be fully up to the average of the Basin.

On the Beaver branch of the Moshannon the measures exposed show but little variation.

Fig. 5. *Beaver Bk*

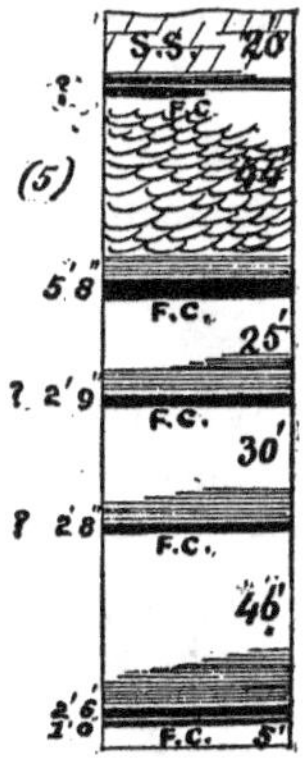

The section Fig. 5 represents the north side of Beaver creek, opposite the Moshannon Mine. The coals are only opened on the outcrop, except D which is opened up for working. This bed shows here nearly 6 feet of clean coal, a very great expansion in size, the average not being more than about 4 feet. Bed B as mined by the Moshannon Coal Company on the south side of the Beaver creek is a fine 5 foot coal bed: where opened on the crop on the north side it shows 3½ feet. Bed D only comes into the extreme hill top on the south side of the creek. It is reported as once opened there 6′ thick.

Fig. 6. *Houtzdale.*

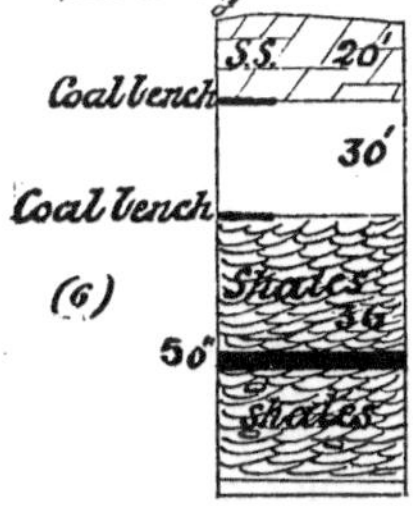

Two miles west south-west of this point the following section is fouud at Houtzdale (Fig. 6.) The measures here retain their average character; Bed B, however, existing with unusual size and purity, showing over 5 feet of clean coal, very free from injurious impurities.

Fig. .7 *Goss Bk*

Section Fig. 7, shows the north-west side of the Goss Hill, where the coal beds come to daylight on the steep hill side. It gives:

Shales at hill top, - - - - -	10′
D. Coal bed, outcrop.	
Concealed measures, - - - -	35
Black slate outcrop.	
Concealed measures, - - - -	90
Black slate outcrop.	
Concealed measures, - - - -	80
Black slate outcrop.	
Sandstone, - - - - - -	15
Bed of Creek.	

The hill side is covered with loose stuff preventing exposure of the measures; and the bed of the small run is filled with massive fine grained sandstone boulders.

On the west side of this run outcrops of slate and defined benches are found at 30′, 135′ and 160′ above water level.

At the Henry Goss Farm, 1 mile north-west of the run, coal is opened 100′ above water level. It is probably Bed A, as a micaceous sandstone shows, overlying, very similar to that observed at Osceola.

From here north-westward, one-half mile to the first anticlinal sub-axis, a fine grained massive conglomeritic sandstone makes up the country rock. In one lump of this massive sandrock was found a small piece of coal, of filbert size, packed in closely with rounded water-worn quartz grains. But the coal itself gave no sign of having been rubbed or worn, and clearly had not traveled far after being torn away from its original place of deposit.

In this First Coal Basin there occur some small faults. True faults must be rare in the bituminous coals; and in the whole season's work not a single one was found outside of the First Basin, which lies close to the great axis of the Allegheny mountain. Even in this basin no fault exceeding 10 feet in upthrow was noticed.

Such slight faults are easily overcome in working; but a much more serious difficulty is the presence of what the miners call "horsebacks," "clay-veins," "troubles," "nips," "wants," etc. By these terms are expressed the various difficulties met with in mining, such as the uneasy and rolling floor and roof of the coal bed; sudden thinning of the coal; or where part of it is rotten and soft, resembling the gob waste of the mine; sandstone, clay or shale in irregular layers in the coal; or sandstone for some distance taking the place of the coal. All of these troubles are met with in Clearfield and Jefferson counties, and are described in the detailed report. They seem more abundant in the First Coal Basin than in those to the west of it.* Moreover, the Upper Coal Beds appear to be less troubled than the lower ones, bed D being more regular than beds A or B.

A striking instance of the short distance within which a

* [Probably because this basin has been more extensively mined. J. P. L.]

smooth and even bed of coal can become twisted, cut out and ruined for mining, is furnished by the Eureka and Penn Collieries at Houtzdale. The properties lie side by side, along a smooth hillside, which gives no outward evidence of any disturbance. And yet, as will be seen by the detailed description, in one of these two collieries the bed is unusually even and regular, while the neighboring collieries working the same bed have been abandoned on account of the continuous rolling and pinching out of the coal.

CHAPTER X.

Description of the Mines in the First or Steam Coal Basin of Clearfield County.

Nearly all the Steam Coal shipped at present from Clearfield county comes from some 12 to 15 collieries lying at intervals between Houtzdale on the Beaver Branch of the Moshannon creek, and the Morrisdale mine, 3 miles north-west of Phillipsburg. Only two coal beds are opened and worked for shipment, (B and D of section Fig. 1.) The Clearfield Steam Coal Basin Map (Plate V) shows the position, of the different mines.

Mines Working Coal Bed B.

The only bed worked along the Beaver Branch of the Moshannon is B, excepting where the Moshannon company have recently opened up one mine on D.

At Penn colliery at Houtzdale, B is opened on the north side of the Beaver Branch, about 25 feet above the Creek level. The main entry goes in north about 700 to 800 yards. For the first 100 yards the roof and floor are comparatively regular, giving in different places from 42″ to 58″ (inches) of coal, hard, black, shining and very clean. It shows:

Fig. 8.
Penn. Bank

Shales.		
Black slate roof, - -		2′ to 4′
Sandstone, persistent, -		8″
Black slate, hard, - -		1 to 2
Coal, - - - - -	3′ 6	to 4 10
Fire-clay floor, - -		4

The coal breaks out in blocks, showing the columnar structure which characterizes all these coals; and bears shipment well.

After the troubles begin, about 100 yards in, the floor sometimes rises and cuts down the thickness of coal, or the roof descends, making only a crushed and useless mass of slate and coal. These troubles continue on with varying intensity to the main heading, and when the mine was examined, (July, 1874,) the proprietors were proposing to rob out and abandon it.

An average specimen shows, by Mr. M'Creath's analysis, in the Laboratory of the Survey at Harrisburg:

Water,	.810
Volatile matter,	20.640
Fixed carbon,	74.023
Sulphur,	.507
Ash,	4.020
	100.000

Coke per cent. 78.550.—Color of Ash, white.

Analysis of the Ash:—

Silica,	2.040
Oxide of Iron,	.350
Alumina,	1.140
Lime,	.136
Magnesia,	.032
Phosphoric acid,	.007
Sulphur,	——
Per cent. of Ash in coal,	4.020

An analysis of coke made at this mine, roughly, in the open air, gives:—

Water at 225°,	600
Volatile matter,	2.020
Fixed carbon,	88.032
Sulphur,	.998
Ash,	8.350
	100.000

Coke yields a Red Ash.

"The coal is deep black, with shining lustre, of a somewhat columnar structure."

At a distance of 2,300 feet (north 15° west) from the main opening of the Penn Mine, a drift has been run in on this bed B. The same difficulties are encountered, the coal cutting down from 40″ to 12″ or less.

The same uneven fire-clay floor and black slate roof show here. In both openings the roof slates are filled with the impressions of the large plants common in the lower part of the Lower Coal Measures, and a very few ferns.

The Franklin Colliery of the Kittanning Coal Company, is opened on the south side of the Beaver Branch at Houtzdale. It lies only about 20 feet above the creek, and the mine runs in about 800 yards to the south 16° east, the coal rising very gently to the south-east.

Fig. 9.

The Superintendent of the mine reports the following section above the open bed:—

Coal, - - - - - - - - - -	1′	8″
Sandstone, - - - - - - - -		8
Coal, - - - - - - - -	10	
Sandstone, - - - - - - - -	5	6
Black slate and thin sandstone layers, -	2	6
Sandstone (visible) persistent, - -	6	
Black slate, - - - - - -	1	6
Coal B., 58″ to 64″, (say 5 feet.)		
Fire-clay floor.		

An apparently clean upthrow of 10′ occurs in the mine at one place, obliging the Company to cut through 10′ of their fire-clay floor, and some calcareous slaty nodules, for a distance of from 150 to 200 yards. The coal, however, was not injured nor cut down; but, where measured in different parts of the mine, ranged from 58 to 64 inches of clear bright coal in one bench, without persistent slate partings.

An average specimen of this coal yielded, on analysis, (M'-Creath):—

Water at 225°, - - - - - -	1.942
Volatile matter, - - - - -	22.720
Fixed carbon, - - - - - -	71.018
Sulphur, - - - - - - -	.543
Ash, - - - - - - - -	3.777
	100.000

Coke per cent, 75.340

"The coal is bituminous, with bright shining lustre and deep black color, somewhat columnar structure, very friable, and containing considerable mineral charcoal, also numerous small scales of Calcite and a small amount of Iron Pyrites.

The coal does not swell much during coking and forms a good coherent coke with a dull metallic lustre and yields a cream colored Ash."

The President of the Kittanning Coal Company has furnished a copy of an analysis of this coal made for the company by Prof. Chas. A. Seely, chemist. The analysis corresponds closely with the general average of the coal of this part of the basin as determined by Mr. M'Creath's analysis. It gives—

"Volatile combustible matter,	20.10
Fixed carbon,	76.39
Ash,	3.51
Coke,	79.09
Sulphur,	0.19

The small percentage of sulphur and of ash, as well as practical tests of the coal by coking and burning in various ways, indicate that for metallurgical purposes and for raising steam, it is of the first quality."

The hill rises very gently above the coal and takes in no other coal beds for some distance back to the south-east. The rise to the bed is so gradual that where it comes out to daylight over on the Moshannon about 1¼ miles from the mouth of the mine it is only 60 feet higher, or less than an average ½° dip.

The Eureka Colliery is opened at Houtzdale on the north side of the Beaver Branch of the Moshannon creek. The main entry runs in north, the coal rising gently to the north-west.

Fig. 10.

Eureka Bk

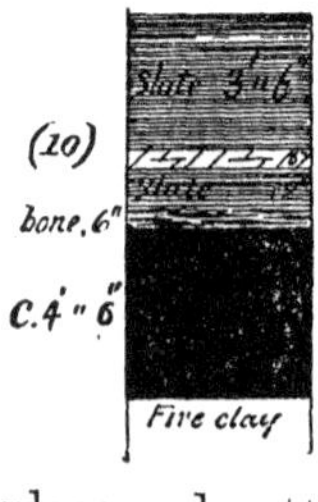

(10)

Slate 3' 6"

Slate 18"

bone, 6"

C. 4" 6

Fire clay

Blackslate,	3′ to 4′
Sandstone,	6″
Black slate,	1 to 2
Bony coal,	6
Coal,	4′ 6″ to 5
Fire-clay floor,	6

The roof and floor are very regular and undisturbed, the roof however rolling down in one place and cutting down the size of the coal for a short distance. There is also an occasional "swamp," but small and easily overcome. An average specimen of the coal yields by analysis, (M Creath):—

"Water,	0.78
Volatile matter,	21.680
Fixed carbon,	73.052
Sulphur,	.688
Ash,	3.800
	100.000

Coke, per cent., 77.540.

"Color of Ash, gray. The coal is shiny black, very soft, with small seams of charcoal."

Analysis of Ash:—

Silica,	1.660
Oxide of Iron,	.560
Alumina,	1.360
Lime,	.134
Magnesia,	.046
Phosphoric Acid,	.013
Sulphur,	——
Per cent. of Ash in coal,	3.800

The coal makes a handsome appearance in the mine, and is without persistent slate partings. The roof is excellent and the coal mines well and bears transportation.

The Company furnishes the following analysis of this coal made by Messrs. Booth and Garrett, chemists, as follows:—

Moisture,	1.15
Volatile matter,	19.50
Fixed carbon,	77.05
Ash,	2.30
	100.00

A section at the Eureka mine is as follows:—

Shales and sandstones, thin,	20′
Small (coal ?) bench.	
Shales and slates, thin and rusty,	25
Slate roof of coal,	5 6″
Coal,	5 6
Fire-clay,	6
Sandstone, brown, rusty, ferruginous,	12
Concealed measures,	6
Railroad and Beaver Branch.	

From the Eureka mine the coal is rising to the north-west, and where opened at Houtzville this Bed B is 100′ (feet) higher, the distance being $1\frac{1}{4}$ to $1\frac{1}{2}$ miles. At the time of the examination (July 1874) the mines of this region were newly opened, and no regular shipments had begun.

At Skeith's, or the Webster Colliery, the coal showed the following section:— *Fig.* 11.

Webster Bank

C. 0" 8
black sl. 1"
C. 4" 8
(11)
Fire clay.

Black slate roof.		
Coal,	- -	0′ 8′
Black slate,	-	1
Coal,	- -	4 8
Fire-clay floor,		8 more or less.

The mine runs in to the north 60° west, and drains, the coal rising gently but steadily to the First anticlinal sub-axis.

A specimen of coal forwarded by the owners to Mr. M'Creath, analyzed—

Water at 225° F., - - - - - -	1.630
Volatile matter, - - - - - -	22.000
Fixed carbon, - - - - - -	72.815
Sulphur, - - - - - - -	.425
Ash - - - - - - - -	3.130
	100.000

Coke per cent., 76.370.

"The coal has a bright resinous lustre, is of somewhat columnar structure, and very friable. It contains numerous veins of bright crystalline coal and mineral charcoal, and shows very little iron pyrites. The coal swells but little during coking, yielding a good coherent coke and gray ash with slight reddish tinge. If this sample fairly represents the mine, it shows a coal of very superior quality, and compares favorably with any yet examined here."

At the Diamond Mine, opened a few hundred yards further north, the coal shows this section:—

Black slate roof.		
Coal and bony coal,	- -	0′ 8″
Slate	- - - - - -	1.5 to 2
Coal,	- - - - -	4 10
Fire-clay floor.		

From the measurements and the analysis, it is clear that coal bed B retains in this Houtzville region the same handsome size and high character displayed by it along the Beaver Branch.

South-west of Houtzdale, Messrs. Kendrick & Co. have been developing the coals. They report thus: "On the south side of the valley, the Moshannon bed (B,) is found about 6 feet thick, with a dip of $\frac{3}{8}°$ to the south-west. There is a synclinal in a swamp, and on the rise of the measures, a shaft was sunk through the following rock measures:

Surface stuff, - - -	18′
Shale, olive, - - -	12
S. S. hard bluish gray, -	34
Black slate, - - -	5
Coal, (with iron pyrites,)	0 8″
Parting.	
Black slate and poor coal,	3
Parting.	
Coal, - - - - -	2 9
Fire-clay floor.	

The above is the description given by the operators themselves.

The Stirling Colliery of the Powelton Coal and Iron Company, is opened on the south side of the Beaver Branch of the Moshannon creek, about one mile north-east from the Franklin colliery, and about 20′ above the creek. The coal shows:

Fig. 12.

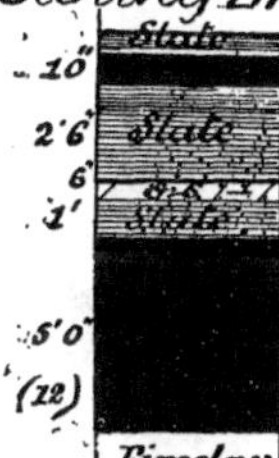

Roof, black slate, - -	2′ 6″
Sandstone persistent, -	6
Black slate, - - -	1
Coal, - - - -	5
Fire-clay floor.	

The coal is in one bench, without regular and persistent slate partings, and is rising gently to the south-east. The roof is good and firm, and the coal mines out well.

The air shaft shows the following measures resting on top of the roof of slate in descending order:

Surface at top of air shaft.	
Small coal bed, - -	1′
Slates, hard gray, - -	20
Sandstone and slate, -	5
Slate, - - - -	2
Sandstone, - - -	1
Slate, - - - -	0 9″
Sandstone, - - -	0 6
Slate, - - - -	4
Coal bed, - - -	0 10

This small coal 42′ above the main bed, was opened up in another place near by where it was under cover, and found to yield from 30″ to 33″ of good coal.

The main bed has a regular roof and floor; the only troubles in the mine being caused by a few small "swamps" easily overcome.

An average specimen of the coal shows by analysis, (M'Creath,)

Water,	0.710
Volatile matter,	23.400
Fixed carbon,	72.218
Sulphur,	0.532
Ash,	3.140
	100.000

Coke per cent., 75.890. Color of ash, gray with red tinge.

"The coal is black, columnar and contains scales of Iron Pyrites."

The Moshannon mine of the Moshannon Lumber and Mining Company, is opened on the south side of the Beaver Branch, about 1½ miles below Stirling, and 3 miles south-west of Osceola.

Fig. 13.

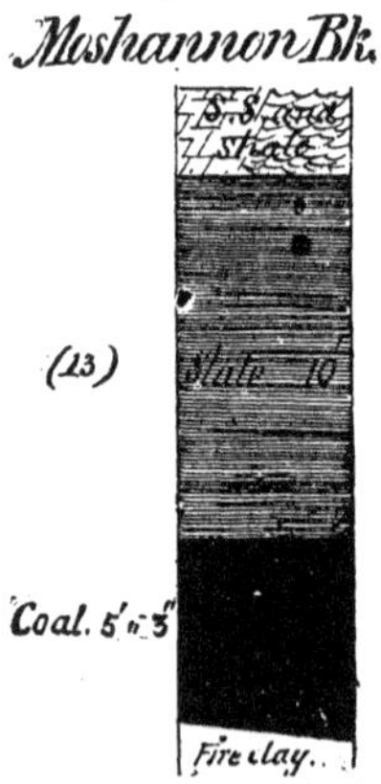

Moshannon Bk.

Roof, black and grayish slates,	10′
Coal,	5′ 3″ to 5 6″
Fire-clay floor,	3 or more.

The coal shows clear and bright in one bench, without any persistent slate partings.

In driving off east from the main entry at about 150 yards distance, a serious trouble was found. The coal dips suddenly downward, say 5 feet in a distance of 30 feet. Then what appear to be the roof slates of the coal, rest directly against the coal itself, in a clean fracture. Some 7 feet below, a 15 inch coal is found with black slate roof and fire-clay floor complete. A shaft from the surface some 15 yards east of the present (July, 1874,) workings, found the same thing, 15 inches of coal. The workings had therefore been carried no farther in that direction than the line of fault. The imperfect data gives the line of fault as east 20° north, and west 20° south, but the direction is by no means certain. The measures are also somewhat disturbed on the south-west side of the property. Going south-east on the surface across the small ravine, on the next hill top to the south-east, 1,000 yards away, the coal bed D is found on the extreme hill top in its proper position, and is reported as having shown, when opened, between 5 and 6 feet of good coal. It is now fallen shut.

An average specimen of the coal from the Moshannon mine (Bed B) yielded on analysis, as follows (M'Creath):

Water, - - - - - - -	.765
Volatile matter, - - - - - -	20.090
Fixed carbon, - - - - - -	74.779
Sulphur, - - - - - - -	.666
Ash, - - - - - - - -	3.700
	100.000

Coke per cent., 79.145. Color of ash, gray with reddish tinge. "The coal is black, shining, friable, contains much iron pyrites and charcoal."

About 35 feet above this coal bed, there is an old opening on a small coal bed, now fallen shut, which is reported as having given 30 inches of good coal.

The Superintendent reports that on sinking a well on the flat just below the mine, (25 feet lower,) a coal was struck 20 feet below the surface, making a coal 45 feet below this Bed B. Coal Bed A is usually from 50 to 60 feet below in most places.

The company have opened up their coal outcrops on the north side of the Beaver Branch, opposite this Moshannon mine. A flat some 200 yards wide separates the hills on the opposite side of the run. This section (Fig. 5,) has already been given. It shows thus:

Hill top.		
Sandstone, - - - - - -	17	
Black slate, - - - - -	3	
Coal, D, - - - - - - -	2	6″ (?)
Concealed measures, - - -	33	
Roof slates, - - - - - -	6	
Coal D, - - - - - - -	5	8
Shales and sandstone, - - -	27	
Coal, - - - - - - -	2	9, (not seen.)
Shales, - - - - - - -	27	
Coal, - - - - - - -	2	8, (not seen.)
Thin sandstones and shales, -	31	
Shales, - - - - - - -	10	
Coal, - - - - - - -	2	6
Slate, - - - - - - -		3 to 6
Coal, - - - - - -	1	
Concealed measures, - -	5	
Beaver Branch creek.		

The coal beds in this section are only opened up on the outcrop, excepting bed D, which was opened well into the solid coal when examined, (July, 1874.) This opening showed:

Fig. 14.
New Moshannon Bk
Slates 6' 
bone 6"
C. 5' 8"
(14)
Fireclay.

Roof slates, gray and black,	-	6'
Black slate and bony coal,	-	0 6"
Coal, - - - - - -	-	5 8
Fire-clay floor, - -	-	2' to 3

The coal where examined was in one bench, without regular and persistent slate partings.

An average specimen of the coal yielded by analysis (M' Creath):

Water, - - - - - -	1.100
Volatile matter, - - - -	23.070
Fixed carbon, - - - - -	71.199
Sulphur, - - - - - -	.611
Ash, - - - - - - -	4.020
	100.000

Coke per cent., 75.830. Color of ash, red.

"The coal has rather a dirty appearance, is friable, and contains considerable Iron Pyrites."

The President of the Moshannon Coal Company, Mr. D. Knight, furnishes the following analysis of this same coal, made by Messrs. Booth and Garrett. The two analyses, it will be seen, agree quite closely:—

Water, - - - - - -	1.100
Volatile matter, - - - - -	22.450
Fixed carbon, - - - - -	72.300
Ash, - - - - - - -	4.150
	100.000

The Beaverton mine which was opened and worked just south-west of the Moshannon mine (on the adjoining tract) is now abandoned. The coal is said to have shown about the same thickness.

The old Decatur mine opened near the head of Coal Run is also abandoned and fallen shut: and affords no opportunity for measurement. It is reported that the mine was faulty and abandoned for that reason.

The next opening on Coal Bed B from which any considerable mining has been done is at the Clearfield Coal Company's Colliery on the south-east side of the Moshannon opposite Osceola. The colliery is not now working.

Fig. 15. The coal as measured in the main entry shows:

Roof, black slate,	3′	1″ or more.
Coal,	2	6
Black slate,	0	6″ to 8,
Coal,	2	not seen well.
Floor		

The upper bench of this coal looks well, hard, black, shining; another drift run in some 300 or 400 yards to the south-west is also abandoned.

The first opening was examined in 1864, by Prof. J. P. Lesley, with the mine well opened and free from water, and thus described:

"This bed is opened in about 100 yards, ready for commencing work. I measured it in two places, forty yards apart, and found it regularly divided into two benches; the upper measuring 2 feet 10 inches, with six inches of slate between them; the lower measuring two feet.

The upper bench yields a beautiful, pure, rich bituminous coal, crystalized vertically, and I should say from its appearance, admirably suited for iron work.

The lower bench yields a harder, more compact coal, and in one place showed a trace of central slate, but will answer well for the same purpose.

The intermediate slate is not sandy, and therefore will not give the mine any trouble; in fact, may be expected to disappear in various parts of the workings.

The bed will yield four feet of good marketable coal at the least, allowing for all variations, and will sometimes yield five feet or more."

The Lower Coal Bed A shows just above water level on the Moshannon below the dam.

The Enterprise Colliery is opened on the south-east bank of the Moshannon creek, ½ mile below Osceola, 35 feet above the water level.

Fig. 16. Enterprise B'k

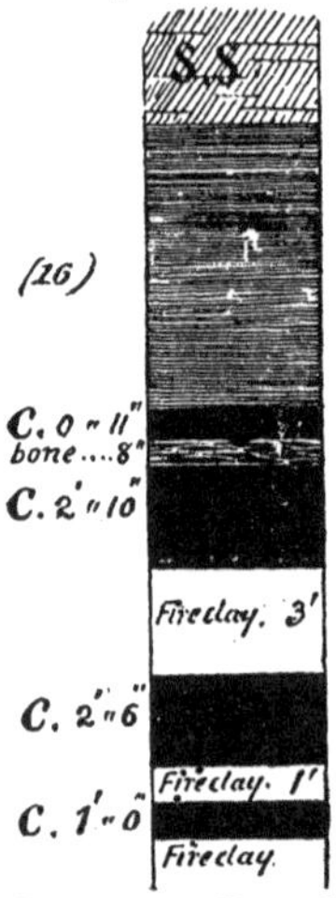

The mine where examined, showed:

Sandstone.		
Top, black slate, - - - -	8′	
Coal, 11″, reported - - -	1	4″ in places.
Bone coal and slate, - - -		8
Coal 2′ 10″, reported - - -	3	4 in places.
Impure fire-clay, - - -	3	
Coal, - - - - - - -	1	or more, bottom not seen on account of water.

Mr. Zeigler, the Superintendent, reports this 12 inch coal as a 30 inch coal, and thus continues the section: coal 30 inches, (instead of 12;) fire-clay, 1 foot; coal, 1 foot; fire-clay bottom; none of this was visible when the mine was examined.

Mr. Zeigler reports, that in his workings to the south-west, he finds the coal sinking in that direction; showing the presence here of one of those small rolls which disturb an average dip, the average sinking here being to the north-east.

The coal looked decidedly sulphurous, iron pyrites showing constantly; no fair average specimens could be obtained for analysis.

A small coal bed is reported in or just below the bed of the Moshannon at this point. It could not be seen.

A section up the hillside here on the south-east side of the Moshannon showed nothing until a point 215 feet above the stream where there is a 14 inch coal, with massive sandstone boulders covering all the ground above.

The Powelton Colliery of the Powelton Coal and Iron Company is opened and worked on the line of the Tyrone and Clearfield Railroad, 3 miles east of Osceola. The mine runs in north-east, passing under the railroad and shows:—

Fig. 17. Powelton B'k

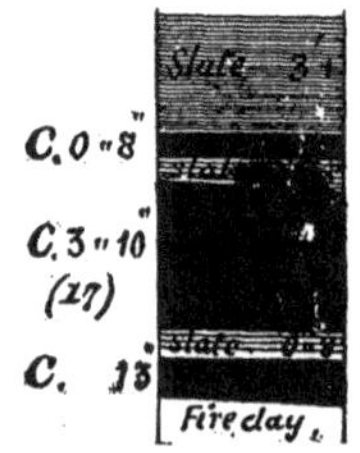

Hard slate roof, - - -	3′	or more.
Coal, poor, not mined, - -	0	8″
Black slate, - - - - -	0	6 to 0′ 9″
Coal, - - - - - - -	3	8 to 3 11
Black slate, - - -		8
Coal, - - - - -	1	to 1′ 2″
Fire-clay floor, - - - -	1	6 or more.

The coal of the main bench is hard, bright, shining coal—

comes out in lumps—somewhat sulphurous, the pyrites showing as specks in the coal and largely in the slate binders. The coal is used on the engines on the Tyrone and Clearfield Railroad and is a strong steam coal.

It is one of the oldest and most extensively worked mines of the region.

A fair average specimen of the coal from the upper or main bench shows, on analysis, (M'Creath):

Water, - - - - - - -	0.540
Volatile matter, - - - - -	22.560
Fixed carbon, - - - - - -	71.551
Sulphur, - - - - - - -	1.079
Ash, - - - - - - - -	4.270
	100.000

Coke, per cent., 76.90. Color of Ash, light gray.

"The coal has a comparatively dull lustre, is hard, with slate and Iron Pyrites through it in veins."

A fair average specimen of the lower bench coal shows, on analysis, (M'Creath):

Water, - - - - - - - -	0.600
Volatile matter, - - - - -	22.600
Fixed carbon, - - - - - -	68.709
Sulphur, - - - - - - -	2.691
Ash, - - - - - - - -	5.400
	100.000

Coke, per cent., 76.80. Color of Ash, gray, with pink tinge.

"The coal is bright, shining, columnar, with small veins of Pyrites and charcoal."

An analysis of the ash from the lower coal bench gives, (M'Creath):

Silica, - - - - - - - -	1.460
Oxide of Iron, - - - - - -	2.480
Alumina, - - - - - - -	1.050
Lime, - - - - - - - -	.180
Magnesia, - - - - - - -	.169
Phosphoric acid, - - - - -	trace.
Sulphur, - - - - - - - -	.082
Per cent. of Ash in coal, - -	5.400

This Powelton Coal Bed (B) outcrops near the Railroad station. The Superintendent reports that a shaft put down back of the station struck the next bed below at 45 feet, ("a 4 foot bed, sulphurous,") Bed A, and nodular iron ore in considerable quantity 10 feet below the coal.

These facts, with the exposures on the hillside at Powelton Mine, give this section of the measures:—

Hill top		
Shales, with their sandstones, -	40′	
Coal, not seen, reported, - -	2	
Shales, nesty, with some ore balls,	30	
Coal, reported, - - - -	3	(block coal.)
Shales, - - - - - -	35	
Coal, B, - - - - - -	5	
Concealed measures, - - -	45	
Coal, A, - - - - - - -	4	
Concealed measures, - - -	10	
Iron ore balls.		

The collieries described above comprise about all of those shipping from this Clearfield Steam Coal Region, working on Coal Bed B.

Mines Working on Coal Bed D.

The larger part of the coal shipped east from the Clearfield region comes from the collieries already described, working Bed B; the balance from those now following, working Bed D.

This bed is opened on the hill top, back of Osceola, and is worked in a small way to supply the village. There are also some old farmer's openings round the neighborhood.

Langdon's Colliery, about 1 mile north-east of Osceola, was not working when the district was examined, and no specimens were taken from it. The coal is run down a tramroad to Dunbar station, on the Tyrone and Clearfield Railroad, for shipment. The mine is opened about 125 feet above the Moshannon, and the hill rises 45 feet above, taking in near the top the small coal bed which everywhere overlies Bed D through this country, about 35 feet above it.

At Hale's Colliery, about ½ mile north of Langdon's Mine, both of these beds are opened. The mines lie about 1 mile north of Osceola, and are on the south side of Shimmel's run. A small branch connects the mine with the Mapleton Branch Railroad, and by that with the Tyrone and Clearfield Railroad.

Fig. 18.

The lower bed shows the following:—

Roof, black slate, tough.
Coal 3′ 8″ to 3′ 10″.
Fire-clay floor.

The black slate roof, though tough and strong, is very irregular and unquiet; coming down frequently and pinching the coal down to 8 or 10 inches of thickness. This irregularity begins almost at the mouth of the mine and continues at intervals up to the end of the main entry.

An average specimen of this coal yields on analysis, (M'Creath):

Water, - - - - - -	0.740
Volatile matter, - - - -	25.210
Fixed carbon, - - - - -	68.628
Sulphur, - - - - - -	2.122
Ash, - - - - - -	3.300
	100.000

Coke per cent., 74.050. Color of Ash, red.

"The coal is bright, shining, columnar, containing veins of Iron Pyrites."

The upper bed is opened 30 feet above, and shows the following section:—

Fig. 19.

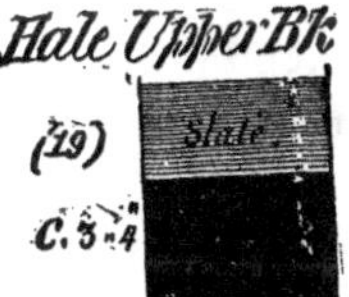

Roof, black slate, crumbly.
Coal, 3′ 2″ to 3′ 6″.
Fire-clay, 1 6 or more.

The roof is poor, but is regular, lacking the rolls and pinches of the bed below.

An average specimen of this coal yields, by analysis, (M'Creath):

Water, - - - - - - -	0.570
Volatile matter, - - - - -	24.630
Fixed carbon, - - - - -	68.400
Sulphur, - - - - - -	1.900
Ash, - - - - - - - -	4.500
	100.000

Coke, per cent., 74.800. Color of Ash, gray, with red tinge.

"The coal is bright, shining, columnar, contains small veins of Iron Pyrites."

Back of the mines, 65 feet by barometer above the upper bed, the surface is covered with lumps and small boulders of hard sandstone, fine grained, slightly conglomeritic in places. This is shown in the section (Fig. 1) which is partly compiled from this place.

Mapleton Colliery is opened on the south side of Shimmel's Run, 1 mile north of Osceola. The mine runs in to the south, the coal rising slowly to the north-west. The coal shows:—

Fig. 20.

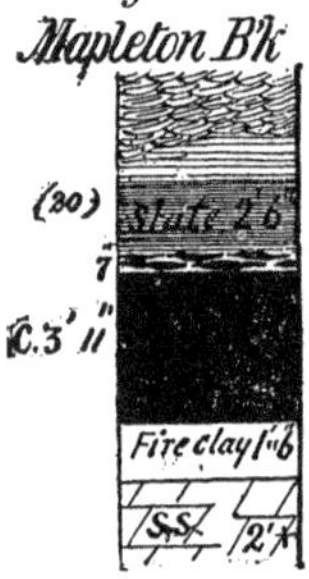

Mapleton B'k

(30) Slate 2'6"

7

C.3' 11"

Fire clay 1'6

S.S. 2'

Grey slates and thin shales.		
Black slate, - - - - - - -	2′	6″
Bony coal, with small coal layers,	0	7½
Coal, - - - - - - - -	3	11
Fire-clay, - - - - - - -	1	6
Sandstone.		

The coal is in one bench, from 3½ to 4 feet thick, without persistent slate partings, and very tender, breaking up easily.

An average specimen shows, by analysis, (M'Creath):

Water at 225°, - - - - -	0.700
Volatile matter, - - - - -	23.565
Fixed carbon, - - - - -	68.890
Sulphur, - - - - - -	1.715
Ash, - - - - - - -	5.130
	100.000

Coke, per cent., 75.735. Does not swell much during coking, forms a coherent coke and yields a gray ash.

An analysis of the ash gives, (M'Creath):

Silica, - - - - - - -	1.675
Oxide of Iron, - - - - -	1.570
Alumina, - - - - - -	1.480
Lime, - - - - - -	.221
Magnesia, - - - - - -	.154
Per cent. of ash in coal, - - -	5.130

The coal is bituminous, with bright shining lustre, columnar structure, very easily broken, contains small scales of Iron Pyrites.

As is usual at most of the mines, some of the small coal had been roughly coked in the open air near the mine mouth. The coking was very imperfectly done: and an average specimen of it yielded by analysis, (M'Creath):

Water at 225°,	.580
Volatile matter,	1.370
Fixed carbon,	84.068
Sulphur,	1.032
Ash,	12.950
	100.000

Has dull lustre, on fresh surface bright, shining lustre, shows much iridescence, and contains considerable slaty matter. Yields a red ash.

The undue proportion of ash in the above analysis shows how carelessly the coking is conducted, and how much bone and slate are put on the hearth with the slack coal.

The mine has an excellent tough black slate roof, and both floor and roof are even and regular.

The bench of the overlying bed shows 35 feet above on the hill side, with about 15 to 20 feet of cover (shales) to hill top.

About 30 feet below the Mapleton Bed (D) is a small coal bed, but not opened so as to be measured.

The Mapleton bed has also been opened on the outcrop on the north side of Shimmel's run, and shows there as well as in the Mapleton Mine.

The Logan Colliery is opened on the north side of Shimmel's run about 1½ miles north of Osceola. The mine runs in about north 5° east, and the coal has a gradual, very gentle dip towards the south-east.

Fig. 21.

Logan Bk.

The coal shows:

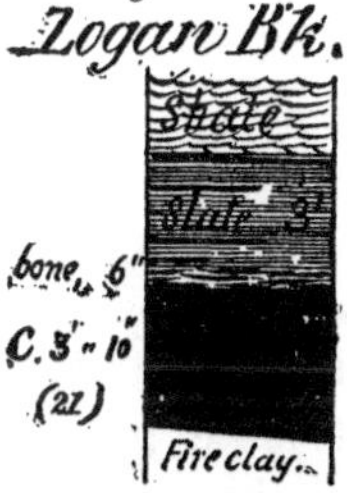

Shales.	
Roof, black slate, tough,	3'
Bony coal,	6"
Coal,	3 10
Fire-clay floor.	

The coal measured 3 feet 8 inches to 4 feet of good bright coal, in one bench, without regular slate partings, usually rather tender.

An average specimen yielded on analysis, (M'Creath):

Water at 225°,	.620
Volatile matter,	22.135

Fixed carbon, - - - - - -	68.728
Sulphur, - - - - - - -	.867
Ash, - - - - - - - -	7.650
	100.000

Coke per cent., 77.245.

The coal is bituminous with dull lustre, somewhat columnar in structure, with thin veins of slate running through it. The coal forms a good coherent coke, with shining lustre. It yields a gray ash.

An analysis of this ash yields, (M'Creath):

Silica, - - - - - -	3.493
Oxide of iron, - - - - -	.750
Alumina, - - - - - -	2.700
Lime, - - - - - - -	.302
Magnesia, - - - - - -	.168
Phosphoric acid, - - - - -	.237
Per cent. of ash in coal, - - -	7.650

Thirty-five feet above the Logan mine, the upper bed is opened up on the outcrop. The section shows:

Hill top.
Shales, - - - - - 15'
Black slate roof.
Coal, D, - - - - - 2 to 2' 6"
Fire-clay floor, bottom not seen.

The Laurel Run Colliery, (Nuttall & Bacon,) is opened on the north side of Shimmel's Run, 1¾ miles north-west of Osceola. The coal shows:

Fig. 22.

Black slate roof.
Bony coal, - - 6" to 12"
Coal, - 3' 10" to 4'
Fire-clay on floor, 3 6, or more.

The coal is in one bench, bright and clean, free from regular and persistent slate partings, and is tender, breaking up easily.

An average specimen of this coal yielded on analysis, (M'-Creath):

Water at 225°, - - - - -	.800
Volatile matter, - - - - -	23.260
Fixed carbon, - - - -	72.350

Sulphur,	.590
Ash,	3.000
	100.000

Coke per cent., 75.940.

The coal is bituminous, with bright shining lustre, columnar structure, containing small scales of Iron Pyrites. It forms a porous, friable coke, with shining lustre. Yields a red ash.

An average specimen of the coke made roughly in the open air near the mouth of the mine, from the coal slack, yielded, (M'Creath):

Water,	0.510
Volatile matter,	1.300
Fixed carbon,	89.243
Sulphur,	.607
Ash,	8.340
	100.000

Color of ash, reddish.

The coke has a silvery lustre, and is very compact.

Reeses' Mine is opened 2½ miles north of Osceola. The coal shows:

Fig. 23.

Roof, black slate and bone coal.		
Coal,		4′ 6″
Black slate,		1 to 3″
Coal,		6
Floor, not seen.		

The coal is bright, clean and hard, and shows very little Iron Pyrites. The roof and floor are hard, good and entirely regular and undisturbed. The mine is only opened for farm use, having no railroad connection, and making no shipments. The coal at this point appears to be at its very best, both in size and quality

The Derby Colliery is opened on the north-west side of the Moshannon creek, opposite Philipsburg. Two coal beds are opened 30 feet apart. The mines run in to the westward, the coal sinking gently to the south-east.

The lower coal opening showed:

Fig. 24.

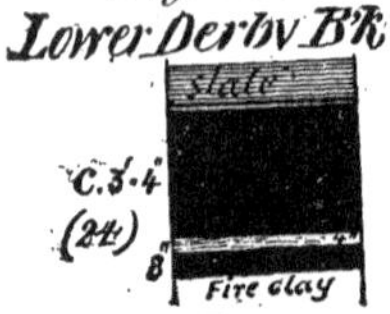
Lower Derby Bk
slate
(24)
Fire clay

Roof, black slate.		
Coal, - - - -	3'	4''
Slate parting, - - -		4
Coal, - - - -		8
Fire-clay floor, - -	1	6 or more.

An average specimen of this coal yielded on analysis, (M'Creath):

"Water, - - - - - -	0.41
Volatile matter, - - - - -	22.81
Fixed carbon, - - - - -	66.69
Sulphur, - - - - - -	1.79
Ash, - - - - - - -	8.30
	100.00

Coke per cent., 76.78. Color of ash, gray with pink tinge.

"The coal is bright, columnar, friable, and contains small veins of charcoal."

Thirty feet of shales and thin sandstone separate this opening from the one on the bed above; nodular calcareous masses showing in the shales, just below the fire-clay floor of the upper bed.

The mine is now fallen shut and cannot be measured; it is thus reported:

Fig. 25.

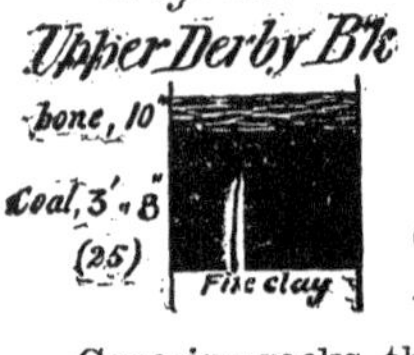
Upper Derby Bk
bone, 10
Coal, 3'.8
(25)
Fire clay

Bony coal, - - -	0'	10''
Coal, - - - -	3	8
Fire-clay floor, - -	2 or more.	

These measures were opened up and worked at the time of the examination for the First Geological Survey of Pennsylvania. The section in Fig. 26 isas follows:—

Covering rocks, thickness not great.		
Coal, - - - - - - - - -	4'	
Fire-clay, - - - - - - - -	2	
Calcareous slate, - - - - - -	8	
Blue compact shale, or indurated clay, -	1	2''
Ferruginous shale, - - - - -	2	
Limestone, blue, ferruginous, - - -	5	
Sandstone and slate, - - - - -	19	
Coal, good. - - - - - - - -	4	4
Fire-clay, - - - - - - - -	2	
Limestone, ferruginous, passing into iron ore,	2	
Sandstone, - - - - - - -	16	
Coal in creek bed, - - - - - -	1	8

Fig. 26.

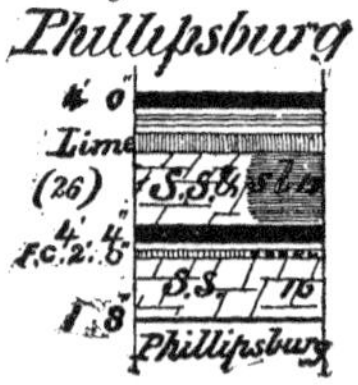
Phillipsburg
Lime
(26)
S.S.
Phillipsburg

The small coal 20 feet below the Derby bed (called 20 inches in Rogers' section) is reported by the mine boss as having been opened up, and shown two small and valueless benches, separated by a 24 inch sandstone parting.

The hill rises nearly 50 feet above the upper opening, and is made up almost entirely of thin shales.

The Decatur Colliery is opened about 1 mile north-west of Philipsburg. The coal shows:

Fig. 27.

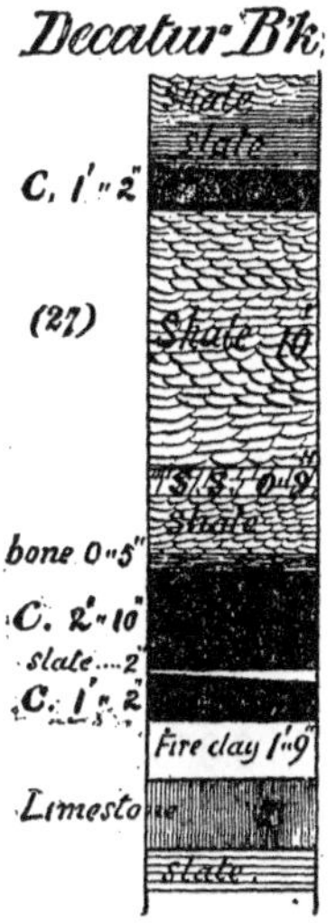

Shale.		
Slate.		
Coal, - - - - - -	1'	2'
Shales, - - - - - -	10	
Sandstone, - - - - -	0	9
Shale and slate, - - -	1	9
Bone coal, - - - - -	0	5
Coal, - - - - - -	2	10
Slate parting, - - -		1 to 4'
Coal, - - - - - -	1	2
Fire-clay, - - 1' 6'' to	2	
Limestone, - - - - -	2	
Sandy slates in bottom.		

The main entry runs in north-west, and the coal dips very gently to the south-east.

The coal is hard, brilliant black, comes out well in blocks, apparently with but little sulphur.

An average specimen of coal from the lower bench yielded on analysis, (M'Creath):

Water, - - - - - - - -	0.640
Volatile matter, - - - - -	24.360
Fixed carbon, - - - - - -	64.082
Sulphur, - - - - - - -	3.378
Ash, - - - - - - - -	7.540
	100.000

Coke per cent., 75.00. Color of ash, gray with pinkish tinge."

The coal has a bright, shining lustre, and columnar structure.

An analysis of the ash gives:

Silica, - - - - - - -	2.100
Oxide of Iron, - - - - - -	3.550
Alumina, - - - - - - -	1.550
Lime, - - - - - - -	.090

Magnesia, - - - - - - -	.206
Phosphoric Acid, - - - - -	tráce.
Sulphur, - - - - - - -	—
Per cent. of ash in coal, - - -	7.540

An average specimen of the coal from the upper bench yielded, on analysis, (M'Creath):

Water, - - - - - - -	0.820
Volatile matter, - - - - - -	23.900
Fixed carbon, - - - - - -	69.007
Sulphur, - - - - - - -	1.373
Ash, - - - - - - - -	4.900
	100.000

Coke, per cent., 75.28. Color of Ash, gray, with pinkish tinge. The coal has a bright, shining lustre and columnar structure."

An average specimen of coke made in the open air from the coal slack yielded, (M'Creath):

Water, - - - - - - -	0.350
Volatile matter, - - - - - -	2.190
Fixed carbon, - - - - - -	90.293
Sulphur, - - - - - - -	.867
Ash, - - - - - - - -	6.300
	100.000

Color of Ash, Red.

The coke is dull gray on outside, silvery on fresh fracture, compact and contains considerable slate.

35 feet above this coal a small bed has been opened up. It is now fallen shut, and is reported as a "30 inch coal," of good quality. Rusty shales, thin, overlie this upper bed for 25 feet to the hill top.

Fig. 28.

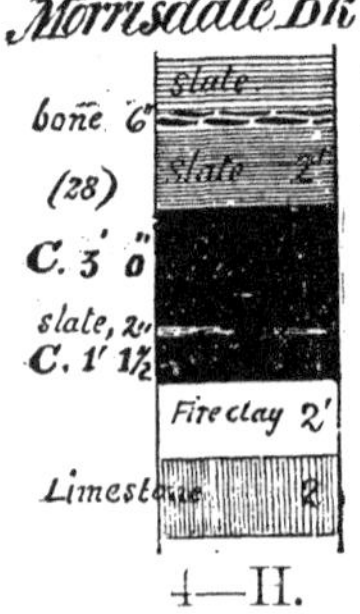

Morrisdale Bk.

The Morrisdale Colliery is opened on both sides of a small branch, and 3 miles north north-west of Philipsburg. The coal shows:—

Greyish tough slate.		
Bony coal, - - -		6'
Hard slate, - -	1' to	3
Coal, - - -		3
Slate, not sandy,	0' 3'' to	0 1''
Coal, - - -		1 1½
Fire-clay, - -		2
Limestone, -		2, ot seen.

The coal from both the upper and lower benches comes out hard, bright and clean. Some few knife edges of slate come in at intervals but none persistent.

The pit boss reports a "5 foot" fault near the south end of the property. He gives the fault a north-west and south-east course. Work was stopped at the fault, and as that part of the mine had fallen shut, it could not be examined.

The mine lies on the Morrisdale Branch Railroad and joins the Tyrone and Clearfield Railroad at Philipsburg.

An average specimen of the lower bench of the coal yields, (M'Creath):

Water,	0.55
Volatile matter,	24.09
Fixed carbon,	71.689
Sulphur,	.571
Ash,	3.10
	100.000

Coke per cent., 75.36. Color of ash, gray.

The coal is bright, shining, columnar, and free from visible Iron Pyrites. An analysis of the ash yielded, (M'Creath):

Silica,	1.450
Oxide of Iron,	.350
Alumina,	.500
Lime,	.260
Magnesia,	.198
Phosphoric acid,	.047
Sulphur,	.054
Per cent. of ash in coal,	3.100

An average specimen of the main upper bench, gave (M'-Creath) on analysis:

Water,	0.56
Volatile matter,	25.19
Fixed carbon,	71.013
Sulphur,	.587
Ash,	2.65
	100.000

Coke per cent., 74.25. Color of ash, salmon.

The coal is bright, shining, contains considerable Iron Pyrites, breaks into plates.

An average specimen of coke, made roughly in the open air, from coal slack, yielded on analysis, (M'Creath):

Water,	0.250
Volatile matter,	0.730
Fixed carbon,	90.707
Sulphur,	.643
Ash,	7.670
	100.000

Color of ash, red.

The coke is coherent, compact, with gray lustre, containing considerable slate.

North-east and north-west of Morrisdale, there has been some opening up of the coals, more for identification and measurement, than for actual present working.

The Clearfield County Map, (Plate IV,) giving the position of the First anticlinal sub-axis, shows that the openings made in Graham township, lie in the Second or western division of the First Coal Basin.

The coal beds have been shafted on or drifted into in three places, all within a moderate distance. The first opening shows:

Black slate roof.		
Coal smut,	1′	
Coal,	1	6″
Shale,	0	7
Coal in sight,	1	6

This lower bench is reported as showing 3′ 6″ more when fully exposed or 5′ in all. This full thickness was not seen, but the authority for the statement is reliable.

The coal lies apparently about 40 feet above the top of the seral conglomerate; a small smut showing on the road side 20 feet below it.

This outcrop opening speaks very favorably for the bed.

One mile south of where Alder run crosses the Grahampton and Kylertown road, a drift has been put in on a coal outcrop. The coal shows the following section:

Roof, bluish slate.		
Coal smut and bone,	0′	6″
Coal,	2	6

Shaly parting, - - - - -	0′ 6″
Coal, - - - - - - - -	1
Fire-clay floor.	

This gives 3½ feet of coal; a good workable bed. Near the mouth of the drift, on the north side, there is a hard micaceous sandstone cutting out some of the coal; thus showing some of that irregularity which is so frequently found in working coal beds A and B.

The third opening is made on the A. Hubler place, in Graham township, not far from the previous trial pits. This Hubler mine is worked in a small way for local use, yielding from 1,000 to 2,000 bushels a year.

The coal measured as follows:—

Slate roof.	
Bony coal, - - - - - -	1′
Coal, - - - - - - - - -	3 3″
Light colored shaly parting, -	1
Coal, - - - - - - - - -	1
Fire-clay floor, - - - - -	0 9

The coal mines out bright and clean.

An average sample forwarded for analysis, yielded (M'Creath):

Water at 225°, - - - - -	0.420
Volatile matter, - - - - -	25.010
Fixed carbon, - - - - -	67.221
Sulphur - - - - - - -	2.479
Ash, - - - - - - - -	4.870
	100.000

Coke per cent., 74.570. Color of ash, Pink.

A section made at this place gives the following series:—

Hill top.	
Concealed measures, - - - - - -	30
Coal, Hubler, - - - - - - - - - -	4 3″
Micaceous sandstone and *concealed measures,*	39
Slate, - - - - - - - - - - -	11
Coal, not seen, called - - - - - -	4
Concealed measures, - - - - - - -	20
Coal, not seen.	

The lower coal is regarded by Mr. Young, who examined these three openings, as resting close upon the Conglomerate.

This region therefore appears to possess Coal Bed A in an excellent condition; and to take in also Coal Bed B as a workable bed.

The region, moreover, is well situated with reference to access to market. From this point the coal measures of this Western sub-division of the First Coal Basin pass on north-eastward to Karthaus.

The section opposite Philipsburg (Fig. 24) reproduced from the Final Report of First Geological Survey, gives the measures as then exposed. "The middle coal bed was at that time mined 4 feet 4 inches thick; of sound texture and excellent quality throughout, affording solid blocks of coal the whole thickness of the bed."

The openings on the John Goss place, 4 miles west south-west of Philipsburg, and 3 miles north-west of Osceola, are now fallen shut and cannot be measured. They were open at the time of the First Geological Survey, and are thus described, (Fig. 29:)

Coarse brown sandstone.
Black slate.
Coal, 7′ too near surface to be wrought.

Concealed measures,	20′	
Coal,	2	
Concealed measures,	15	
Coal,	6	2″
Slate,		8
Coal,	2	

Limestone, thickness unknown.
Iron ore.

Fig. 29.

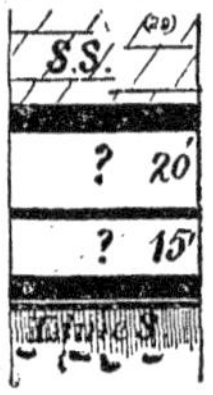

Fig. 30.

"The lowest coal seam being separated into two beds, only the uppermost, which is 6 feet thick, is worked. It affords an excellent sound coal, free from sulphur and slate."

The Pennsylvania Final Report (Hodge) also mentions that coal beds A and B were opened up many years ago on the dividing ridge between Cold Stream and Trout Run and about 1½ miles north-west of the Mountain Summit, (section of Coal Hill, Fig. 30.)

"The lowest bed (A) was found to contain 5½ feet of good coal resting on more than 6 feet of excellent fire-clay: the fire-clay being only 6 to 8 feet above the top of the seral conglomerate.

Fifty feet above this Bed occurs another, 9 feet thick (see section) divided in two seams, by 10 inches of interposed slate, 2 feet from the bottom (bed B.) The upper division, 6 feet thick is alone worked. The coal is of sounder texture than most of that near the mountain admitting of being mined in large square blocks. It is free from sulphuret of iron. The same bed has been opened 2 miles nearer to Philipsburg.

None of these openings now show, and the above section and description are therefore reproduced.

CHAPTER XI.

Special description of Beds A, C and D′, in the First Basin, Clearfield County.

It has already been stated that the coal now shipped from Clearfield county, and known as Clearfield Steam Coal, comes from the collieries opened on coal beds B and D. The sections given with the detailed report on these beds also show coal beds A, C and D′ above water level.

Though these latter beds are not worked for shipment, yet they have been sufficiently opened by trial pits and for local use to afford a fair idea of their general condition and value to the Basin.

Coal Bed A was opened up a number of years ago, just above the water level of the Moshannon creek, opposite Osceola. The mine is now fallen shut. It is 62 feet below bed B, and separated from it mainly by massive conglomeritic sandstones. The opening was examined by Prof. J. P. Lesley, in 1864, and described as follows:

"The heading of the gangway being under water, because it had been driven slightly downward through the roof-rock, in order to facilitate a bridge crossing of the creek at its mouth, I could only measure and examine the upper part of the bed, under a compact clay slate roof of two feet, making an excellent roof for mine work. I saw three feet of hard coal, and have no doubt that the alleged thickness of four feet nine inches (4′ 9″) at the extreme heading is correct. The coal, as I saw it, was in a compact bed without partings, except that an inch of slate came and went in places near the top slate; and in one place I saw a similar thin lens of slate half-way down. I should describe the coal as compact, portable, inclined to ash, and somewhat sulphurous, weathering in peacock colors, a strong burning coal, excellent for steam use and domestic purposes."

Another opening on the same bed near Osceola is reported to have found the bed about as described above; but it was very sulphurous and the working was abandoned.

About ½ mile up Trout run, near the saw mill, the same bed was opened up many years ago. It is now abandoned and fallen shut. The coal is described as having the full thickness of nearly five feet, but was too sulphurous for shipment.

The Powelton Coal and Iron Company shafted on to the same bed just back of Powelton station, on the Tyrone and Clearfield railroad. They found it a strong solid coal of full thickness, but again too sulphurous for use. The same company, as reported by the Pit Boss, opened the bed south-east of the station, about where the bed comes to daylight, and found it in the same condition.

A New York Company opened what is probably the same bed some 1½ to 2 miles east of Philipsburg, on a branch of Williamson's Run. The mine is now abandoned and fallen shut.

The Superintendent states that it showed:

Sandstone roof.	
Coal, - - - - - - -	4′ 4″
Fire-clay floor.	

The coal was here also very sulphurous.

The Superintendent describes the upper bed (B) opened in the same hill (now fallen shut), 60 feet above, as showing:

Black slate roof.	
Coal, - - - - - - -	0′ 6″
Small slate parting.	
Coal, - - - - - - -	2 9
Soft fire-clay, - - - - -	1 6
Coal, - - - - -	4′ 4″ to 4 6
Fire-clay floor.	

At Williamson's mine, on Williamson's creek, 1 mile east of Philipsburg, the coal exposed shows

Slate roof, - - - - . -	1′
Bony coal, - - - - - -	0 2′
Coal, - - - - - - -	1 1
Slate, - - - - - - -	2½″ to 3½″
Floor fire-clay.	

An average specimen of this coal yielded, on analysis, (M'-Creath):

Water, - - - - - - - -	0.62
Volatile matter, - - - - - -	22.73
Fixed carbon, - - - - - -	68.794
Sulphur, - - - - - - -	1.576
Ash, - - - - - - - -	6.28
	100.000

Coke per cent., 76.65. Color of ash, Gray.

The coal is shining, columnar, contains veins of charcoal and pyrites, very heavy.

The sulphur in this Coal Bed A occurs as specks of pyrites in the coal, as masses of Pyrites in the "binders" through the coal, and as huge "sulphur balls," great masses of Pyrites weighing many pounds. It is said that, in past years, numerous openings have been made upon this bed on the branches of the creeks south-east of the Moshannon, but that in all cases the coal was found too sulphurous for shipment.

So far as tried, therefore, this Bed A, south-east of the Moshannon, has proved worthless for shipment; and although the openings and shaftings on to it are not numerous, yet they are sufficient to render it probable that there is but little chance of its proving to be a valuable bed south-east of the Moshannon creek, between Philipsburg and its coming to day-light near Sandy Ridge.

Coal Bed C, 42 to 45 feet, on the average, above bed B, is opened at only a few places in the basin. At the Stirling Colliery, opened 43 feet above bed B, it is said to have given 30 to 36 inches of excellent coal.

At the Moshannon Colliery this bed has been opened up on both sides of the Beaver Branch of the Moshannon; it is called 30 inches to 36 inches of good coal.

This same bed, near Osceola, was examined by Prof. J. P. Lesley in 1864, and thus described:

"Bed C is a 2 foot bed, opened on the run on the road to Philipsburg, at the north-east limit of the property. * * * I consider the bed worthless for mining purposes."

The above conclusion is borne out by the testimony of the

sections given in the detailed report. The bed may locally run 3 feet of good coal; but it cannot be depended on for that amount on an average. The coal is always described as of excellent character.

When a coal basin possesses one or two large and good coal beds, the remaining and less valuable coal beds are entirely neglected; so that there is not now a single complete opening on to this bed in the whole basin, where it can be examined and measured.

Coal Bed D′ is opened at Hale's Colliery, on Shimmel's run, 30 feet above bed D. It gives 3 feet 2 inches to 3 feet 6 inches of good coal in one bench, with a poor, crumbly slate roof and fire-clay floor.

It is also opened above the Logan Colliery at the outcrop, and gives indications of being 24 to 30 inches of coal, rusty shales overlying.

At the Derby Colliery this same bed gives 3 feet 8 incnes of coal, reported of good character.

At the Decatur Colliery, on the Morrisdale Branch Railroad, the bed has been opened up 35 feet above bed D, and is reported as giving 30 inches of coal, with rusty ferruginous shales overlying.

Near the Morrisdale Mine, 35 feet above Coal Bed D, this bed D′ is reported to have given "30 inches of good coal."

For the greater part of the basin, therefore, (where it shows,) this bed is not of workable size; but it seems to be of good character, and in places thickens up to a size which will render mining profitable.

CHAPTER XII.

Summary of thickness and character of Coal Beds in the First Basin.

The detailed description of the collieries has already given in full the main features of the size and character of the different coal beds in this part of the First Coal Basin. These points can be briefly summarized.

Coal Bed A is sufficiently opened to show that it exists through or under almost the entire basin as a 4 foot or 5 foot bed of strong burning steam coal; but that on the south-east side of the Moshannon creek, and where opened at Osceola, north-west of the Moshannon, the amount of sulphur contained is too great to allow the coal bed to be worked for those purposes to which this Clearfield Coal is applied. The numerous trial openings, made so far apart and all agreeing in the presence of iron pyrites in quantities, render it highly doubtful whether over the whole south-east side of the basin, the coal is likely to be found sufficiently free from impurities to be shipped to market.

But where this bed has been opened up for local use on the north-west side of the basin, close to the First Anticlinal sub-axis, the iron pyrites show much less decidedly. It is on the north-west side of this part of the basin, therefore, that this bed offers its best chances.

The measurements and analyses of Coal Bed B, as already given show it to be, on the Beaver Branch of the Moshannon, a very handsome 5 to 6 foot bed of fine steam coal, troubled with horsebacks locally, but on the whole an extremely valuable bed. But its character is by no means of this description throughout the whole extent of the basin. Where opened at Osceola, it showed as a parted bed and sulphurous, and the same impurities are plainly shown in the openings north-east of Osceola. Moreover, the bed has been tried in the region south-east of the Moshannon, between Trout run and Cold Stream, (called Coal Hill in Mr. Rogers' Final Report.) but has always been rejected as not being a first class coal.

The appearance of the parting in the bed at Osceola, and again ½ mile down the Moshannon, with the record of the old opening back on Coal Hill towards the mountain crest, and the description of the abandoned mine 2 miles south-east of Philipsburg, these facts, with the analyses given, would indicate that the normal condition of this bed through the region south-east of the Moshannon, in this part of the basin, is that of a parted and sulphurous coal bed.

The parting is smaller at Powelton Colliery, but the coal carries considerable pyrites.

The very slight opening made on the north-west side of the basin is rather favorable. A very imperfect description of a bore hole at Philipsburg describes the bed as having been passed through, over 6 feet thick.

Where opened recently one mile north of Houtzdale, at the Diamond mine and others, the coal is in most excellent condition, giving its full thickness of 4½ to 5 feet of good coal.

Coal Bed C may be briefly dismissed. It has never been opened as a bed of sufficient size to compete with the other beds in producing coal for shipping to market. It may, and probably will, in places, and for moderate distances, increase to a workable coal bed, say 3½ to 4 feet thick. It is shown in many of the vertical sections and may be called an average 2 foot to 2½ foot bed.

Coal Bed D is extensively worked: and the measurements and analyses already given in Chap. III show it as a regular 4 foot bed, increasing in places to a greater thickness. The coal yielded is an excellent pure steam coal, very even in character. Between Osceola and Morrisdale this bed D is the one worked for shipment. It occasionally shows irregularities of roof and floor, but to a less extent than bed B.

Coal Bed D′ is opened in only a few places and never shows more than 30 to 33 inches of coal, though this coal is usually of excellent quality. It is so persistent in its average size as to afford little prospect of its becoming of any market value as a bed to supply shipping coal from this basin.

CHAPTER XIII.

General Remarks on the First Basin.

The sketch map of the First Coal Basin and the cross section show how the upper beds D and D′ occupy only a comparatively small region in the centre of the basin; going in on their south-east outcrop just north of Osceola, and coming to daylight on their north-west outcrop on Goss Hill. This is but a narrow belt, some 3 miles wide. The map also shows Coal Bed D in the hills south-east of the Moshannon creek; but it is well up in the hill, must have much crop coal, and covers a very moderate area. But as the basin is sinking to the north this coal bed D, which is high in the hill tops just north of Osceola, comes down, until opposite Philipsburg it is low enough to spread broadly out through and under the hills.

The extreme north-west outcrop of the coal, as it rises to the First Anticlinal sub-axis, is apparently about 1 mile or 1½ miles north-west of the Morrisdale Mine. The basin, moreover, ceases to sink to the north, but on the contrary rises in that direction; and the upper bed therefore must shoot out into the air north-east of Morrisdale. The broad flat at Philipsburg practically cuts out Beds D and D′ from the south-east side of the Moshannon creek, at that place. Of course, for these upper beds, lying usually high up in the hills, the erosion has been very heavy; and of the area included between their general outer limits great sheets have been swept off. Of the district mentioned and included on the map, probably one-third to one-half of the workable coal of these upper beds has been washed away.

But Coal Beds A and B cover, except in the valleys of the streams, the greater part of this division of the First Basin, from south-west of Houtzdale to north-east of Morrisdale, and from Houtzdale on to the south-west into Cambria county. Coming in as these beds do on the south-east outcrop between Sandy Ridge and Powelton, they pass through the hills, showing in the different stream valleys, until they come out to daylight in the deep valley on the north-west side of Goss Hill. The lower bed goes in again into the hill on the north-west side of this valley and rises out again into the air to cross over the First Sub-Axis.

Where Bed B is mined on the Beaver Branch of the Moshannon creek, and all along the Branch, nearly down to Osceola on the main stream, the coal is only from 20 to 30 feet above water level. Lying thus low in the hills, it passes through and under all the hills in a broad and regular sheet, always readily accessible, giving a great extent of valuable coal area.

North of Houtzdale it rises regularly to the First Sub-axis; the mines at Houtzville (Webster Colliery) being 115 feet above the Houtzdale mines and only 1 to 1½ miles distant. South of the Beaver Branch the coal rises more slowly to the Allegheny Mountain crest, coming up only 60 feet in the mile which the coal travels from the mouth of the Franklin Colliery to where it comes out to daylight on the west bank of the Upper Moshannon creek. It was not possible to obtain an accurate record of the old boring at Philipsburg, but the coal was reported as a "6 foot bed."

To the south-west of the limit shown on the map, the measures holding these beds A and B extend on continuously to the Maryland line.

The measures to the north-east of Morrisdale have already been described in Chapter X of this report.

CHAPTER XIV.

Character and uses of the Coal from the First Basin in Clearfield County.

The coal shipped east to market from the Clearfield Steam Coal Basin is widely and favorably known, and there is already an extended and growing demand for the coal, on account of its high steam generating power and its value in working iron. Besides supplying large quantities to iron and steel works on the Susquehanna and Schuylkill, much coal reaches tide water at Greenwich Point and Amboy. Some idea of the stability of the trade and its growing hold upon the market, may be inferred from the fact that in 1874, a year of almost unparalleled business depression, the shipment of Clearfield coal amounted to 658,315 tons. In 1873, it was 620,300, showing a gain of 38,000 tons for 1874. The basin is well located, geographically; the coal lies favorably for economical mining, and the production can at any time be largely and suddenly increased to supply any extra demand.

Though with only some ten years of reputation upon them, these coals now rank in the market with the best known and esteemed iron and steam coals; and the analyses given in the body of this report, with the description of the mines, show the average excellent character of the coals, both from the mines working Coal Bed B and those working Coal Bed D.

It may be noted here that these coals are frequently, perhaps usually, termed semi-bituminous coals. They are truly bituminous, having over 20 per cent. of volatile matters on the average, while the term semi-bituminous belongs to the Cumberland, Towanda, Blossburg, &c., coals, which average between 15 and 20 per cent. of volatile matters.*

In order to facilitate comparisons of the analyses of these coals with analyses and power of other standard coals, the analyses of the coals, cokes and ash, of the mines described in the preceding report and that on Clearfield county, are taken from the tables prepared by Mr. M'Creath, Chemical Assistant of the Geological Survey, and grouped together in convenient form, followed by a condensation of Johnson's tables, showing the relation of steam raising power to chemical composition.

Clearfield County Coals.

No.	Water.	Volatile Matter.	Fixed Carbon.	Sulphur.	Ash.	Color of Ash.	Coke, Per cent.
1....	.81	20.640	74.023	.507	4.02	White....................	78.550
2....	1.942	22.720	71.018	.543	3.777	Cream	75.340
3....	.78	21.680	73.052	.688	3.80	Gray	77.540
4....	.71	23.400	72.218	.532	3.14	Gray, with red tinge	75.890
5....	.765	20.090	74.779	.666	3.70	Gray, with red tinge	79.145
6....	1.10	23.070	71.199	.611	4.02	Red	75.830
7....	1.10	22.450	72.300	....	4.15	....	....
8....	.57	24.630	68.40	1.900	4.50	Gray, with red tinge	74.800
9....	.74	25.210	68.628	2.122	3.30	Red	74.050
10....	.70	23.565	68.89	1.715	5.13	Gray	75.735
11....	.62	22.135	68.728	.867	7.65	Gray	77.245
12....	.80	23.260	72.35	.590	3.00	Red	75.940
13....	.64	24.36	64.082	3.378	7.54	Gray, with red tinge	75.00
14....	.82	23.90	69.007	1.373	4.90	Gray, with red tinge	75.28
15....	.55	24.09	71.689	.571	3.10	Gray	75.36
16....	.56	25.19	71.013	.587	2.65	Salmon......	74.25
17....	.41	22.81	66.69	.179	8.30	Gray, with red tinge	76.78
18....	.63	24.63	70.396	.654	3.69	Red	74.74
19....	.75	19.57	69.833	.677	9.17	Gray, with red tinge	79.68
20....	.38	22.28	67.995	2.455	6.89	Dirty gray, red tinge....	77.34
21....	.41	21.80	72.903	1.087	3.89	Red	77.79
22....	.55	22.65	72.616	1.334	2.85	Red	76.80
23...	.48	22.32	59.788	4.232	13.18	Purplish	77.20
24....	.64	23.01	71.799	.551	4.00	Red	76.35
25....	.70	24.02	64.951	1.639	8.69	Red	75.28

Centre County Coals.

No.	Water.	Volatile Matter.	Fixed Carbon.	Sulphur.	Ash.	Color of Ash.	Coke, Per cent.
1.....	.62	22.73	68.794	1.576	6.28	Gray	76.65
2.....	.60	22.60	68.709	2.601	5.40	Gray, with pink tinge...	76.80
3.....	.54	22.56	71.551	1.079	4.27	Light gray	76.90

Jefferson County Coals.

No.	Water.	Volatile Matter.	Fixed Carbon.	Sulphur.	Ash.	Color of Ash.	Coke, Per cent.
1.....	1.01	27.79	48.365	3.885	18.95	Gray, with pink tinge...	71.20

*The nomenclature used in the Final Report of the First Pennsylvania Geological Survey, is as follows:

		Vol. matter.
Anthracites,	Hard Anthracite, - - - - - - -	2 per cent.
	Semi or Gaseous Anthracite, - - - - -	10 "
Common Bituminous, or Coke Coals.	Semi-bituminous, - - -	12 to 18 "
	Bituminous, - - - - -	18 to 48 "
Hydrogenous yielding no Coke.	Cannel coal. Hydrogenous shaly coal, - - - Asphaltic coal.	30 to 70 "

Name and Locality of Clearfield County Collieries.

No.

1. Penn Colliery, Houtzdale, 5½ miles S. W. of Osceola.
2. Franklin Colliery, Houtzdale, 5½ miles S. W. of Osceola.
3. Eureka Mine, Houtzdale, 5½ miles S. W. of Osceola.
4. Sterling Mine, Houtzdale, 5½ miles S. W. of Osceola.
5. Moshannon Colliery, on Beaver Branch of Moshannon, 3½ miles S. W. of Osceola.
6. New Moshannon Mine, north side of Beaver Branch of Moshannon, 3½ miles S. W. of Osceola.
7. New Moshannon Mine. Analysed by Booth & Garrett.
8. Hale's Colliery, 1 mile N. of Osceola. Upper Bed.
9. Hale's Colliery, 1 mile N. of Osceola. Lower Bed.
10. Mapleton Colliery, on Shimmel's run, 1½ miles N. of Osceola.
11. Logan Colliery, on Shimmel's run, 2 miles N. W. of Osceola.
12. Laurel Run Colliery, on Shimmel's run, 2 miles N. N. W. of Osceola.
13. Decatur Coal Co.'s Colliery, ½ mile N. of Philipsburg, Centre Co. Lower Bench.
14. Decatur Coal Co.'s Colliery, ½ mile N. of Philipsburg, Centre Co. Upper Bench.
15. Morrisdale Mine, 3 miles N. N. W. of Philipsburg. Lower Bench.
16. Morrisdale Mine, 3 miles N. N. W. of Philipsburg. Upper Bench.
17. Derby Colliery, ½ mile W. of Philipsburg.
18. Reiter's Coal Mine, near Karthaus P. O. Upper Bed.
19. Mon's Mine, ½ mile S. of Kylertown.
20. Hill's Mine, ⅔ mile E. of Clearfield.
21. Humphrey's Mine, 1 mile W. of Clearfield.
22. Mason's Mine, 1½ miles W. of Clearfield. Upper Bench.
23. Mason's Mine, 1½ miles W. of Clearfield. Lower Bench.
24. G. W. Davis Mine, 1½ miles S. E. of Janesville.
25. Jeremiah Cooper Mine, 1 mile S. E. of Glen Hope, Clearfield creek.

Centre County Collieries.

1. Williamson's Mine, on Williamson's run, 1 mile E. of Philipsburg.
2. Powelton Mine, 3 miles S. E. of Osceola, Clearfield Co. Lower part of Bed.

3. Powelton Mine, 3 miles S. E. of Osceola. Upper part of Bed.

Jefferson County Coals.

1. Brown's Coal Mine, 4 miles S. E. of Reynoldsville.

Analyses of Cokes.

No.	Water.	Volatile Matter.	Fixed Carbon.	Sulphur.	Ash.	Color of Ash.
1............	.600	2.020	88.032	.998	8.350	Red.
2............	.580	1.370	84.068	1.032	12.950	Red.
3............	.510	1.300	89.243	.607	8.340	Reddish.
4............	.350	2.190	90.293	.867	6.300	Red.
5............	.250	.730	90.707	.643	7.670	Red.

Name and Locality of Collieries, (Analyses of Cokes.)

No.

1. Clearfield Co. Penn Colliery, Houtzdale, 5½ miles S. W. of Osceola. Coked in open air roughly.
2. Clearfield Co. Mapleton Colliery, on Shimmel's run, 1½ miles N. of Osceola. Coked in open air roughly.
3. Clearfield Co. Laurel Run Colliery, on Shimmel's run, 2 miles N. N. W. of Osceola. Coked in open air.
4. Clearfield Co. Decatur Coal Co.'s Colliery, 1½ miles N. of Philipsburg. Coked in open air.
5. Clearfield Co. Morrisdale Mine, 3 miles N. N. W. of Philipsburg. Coked in open air.

Analyses of Ash of Coal.

No.	Silica.	Oxide of Iron.	Alumina.	Lime.	Magnesia.	Phosphoric Acid.	Sulphur.	Per cent. of Ash in Coal.
1 ...	2.040	.350	1.140	.136	.032	.007	....	4.020
2 ...	1.660	.560	1.360	.134	.046	.013	....	3.800
3....	1.675	1.570	1.480	.221	.154	.013	....	5.130
4....	3.493	.750	2.700	.302	.168	.237	....	7.650
5....	2.100	3.550	1.550	.090	.206	Trace.	....	7.540
6....	1.450	.350	.500	.260	.198	.047	.054	3.100
7....	1.460	2.480	1.050	.180	.169	Trace.	.082	5.400

Name and Locality of Collieries, (Analyses of Ash of Coals.)

No.

1. Penn Colliery, Houtzdale.
2. Eureka Mine, Houtzdale.
3. Mapleton Colliery, 1½ miles N. of Osceola.
4. Logan Colliery, 2 miles N. N. W. of Osceola.
5. Decatur Coal Co.'s Colliery, 1½ miles N. of Philipsburg.
6. Morrisdale Mine, 3 miles N. N. W. of Philipsburg.
7. Powelton Mine, 3 miles S. E. of Osceola. Lower part of Bed.

The following table is condensed from Prof. Johnson's report to the U. S. Navy Department, on the steam raising power of the coals of the United States. As the analyses of the coals tested by him are given, a rough estimate of character can be made by comparing the tables of analyses of Clearfield coals already given:

Designation of Coals.	Steam, in pounds, corrected for temperature of water in cistern, to one of fuel, from 212°	Composition in 100 parts.						Ratio of fixed to volatile combustible matter
		Moisture determined by steam drying apparatus	Volatile matter other than moisture	Sulphur	Fixed carbon	Coke	Earthy matter	
Cumberland, Md., coals:								
New York and Maryland Mining Co.	9.777	1.785	12.309	....	73.503	85.906	12.403	5.971
Neff's	9.442	2.455	12.675	....	74.527	84.870	10.343	5.880
Easby's "coal in store,"	10.018	0.669	14.984	....	76.264	84.347	8.083	5.089
Atkinson and Templeman's	10.699	0.446	15.532	....	76.688	84.022	7.334	4.937
Easby and Smith	9.965	0.893	15.522	....	74.289	83.585	9.296	4.786
"Cumberland" (navy yard,)	....	3.125	14.168	0.714	68.438	85.829	14.983	5.000
Pennsylvania (bituminous) coals:								
Dauphin and Susquehanna	9.343	0.446	13.547	0.269	74.244	85.738	11.494	5.374
Blossburg	9.724	1.339	13.927	0.853	73.108	83.881	10.773	4.946
Lycoming creek	8.911	0.670	13.807	0.030	71.532	85.493	13.961	5.181
Quin's run	10.272	0.836	17.868	0.102	72.787	81.193	8.406	4.046
Karthaus	9.091	1.282	17.948	1.580	73.770	80.770	7.000	4.110
Cambria Co.	9.240	2.455	19.019	1.500	69.373	78.526	9.153	3.656

Mr. A. S. M'Creath, Chemical Assistant of the Survey, has devoted attention to the consideration of some questions which are of practical importance to all who use these coals in working iron. These questions are:

1. Amount of Phosphoric Acid in coal and ash.

2. Condition of the Sulphur in the coal.

3. The amount of loss of volatile matter when stored, tests being made of coal under cover, and of coal exposed.

There has not been sufficient time as yet to gather the mass of facts needed for a proper discussion of these points; and the table of Phosphoric Acid percentages, and proportions of Iron and Sulphur in the coals, are reproduced below without comment:

Analyses of Coals for Phosphoric Acid.

NAME OF COLLIERY.	Per cent. in Coal.	Per cent. in Ash.
Penn Colliery, Houtzdale	.007	.174
Franklin Colliery, Houtzdale	.005	.047
Eureka Mine, Houtzdale	.013	.342
Sterling Mine, Houtzdale	.005	.159
Moshannon Colliery	.006	.162
New Moshannon Mine	.005	.124
Mapleton Colliery	.013	.253
Logan Colliery	.237	3.098
Laurel Run Colliery	.011	.366
Decatur Coal Co.'s Colliery	Trace.	....
Morrisdale Mine, Lower Bench	.047	1.516
Morrisdale Mine, Upper Bench	.022	.830
Derby Colliery	.033	.397
Powelton Mine	Trace.	....
Wm. Holtz's Mine, Lower hard part	.013	.273
Snow Shoe Mine, No. 4, Lower Bed	.020	.191
Webster Mine	Trace.	....

Iron and Sulphur in Coals.

NAME AND LOCATION OF COLLIERY.	Per cent. of Sulphur.	Per cent. of Iron....	Sulphur required by Iron to form Iron Pyrites...........	Sulphur left in Coke.	Per cent. of Sulphur in Coke...........
Decatur Coal Co.'s Colliery	1.373	.595	.680	.842	1.118
Penn Colliery, Houtzdale	.507	.245	.280	.264	.336
Franklin Colliery, Houtzdale	.875	.581	.664	.328	.414
Eureka Mine, Houtzdale	.688	.392	.448	.451	.581
Mapleton Colliery	1.715	1.099	1.256	.568	.750
Logan Colliery	.867	.525	.600	.628	.813
Decatur Coal Co.'s Colliery	3.378	2.485	2.840	....	....
Morrisdale Mine, near Philipsburg	.571	.245	.280	.302	.400
Powelton Mine	2.691	1.488	1.700	....	....
Seley's Bank	.736	.154	.176	....	....
Hoover's Mine, (Ohio Co.)	.726	.294	.336	....	....
Brown's Mine, 4 miles S. E. of Reynoldsville	3.885	3.395	3.880	2.220	3.118
Mason's Mine, 1½ miles W. of Clearfield	4.232	3.780	4.320	3.140	4.067
Hum's Mine, 1½ miles N. W. of Punxatawney	.848	.434	.496	.390	.613
Weber's Mine, 2 miles N. W. of Punxatawney	2.042	1.428	1.632	1.768	2.687
Mongold's Mine, 4 miles S. E. of Troutville	2.288	1.120	1.280	.920	1.362
P. Galusha's Mine, 2½ miles N. W. of Brockway'e	7.611	3.570	4.080	3.456	5.498
Webster Mine, 5 miles S. W. of Osceola	.425	.189	.216	.271	.354

As to the effect of weather waste on these coals, there has not yet been sufficient time to allow Mr. M'Creath to make the continued and long protracted investigations which are needed.

The German Railway Association had certain quantities of different coals exposed for twelve months, and re-examined, when the following losses were determined:

	Weight per cent.	Caloric per cent.	Yield of Coke per cent.
Pease's West Hartley, coking	0.0	0.0	0.0
Glucksburg seam, Ibbenburen	1.4	6.0	4.6
Carl mine, near Dortmund	....	2.6	2.1
Hibernia Mine, Gelsenkirchen	0.4	0.6	2.1
Constantin Mine, Bochum	0.4	0.4	0.0
Borglohe Mine, Osnabruck	2.0	6.0	0 5

These figures would prove that the losses which were sustained in weight, caloric power and yield of coke, though appreciable after one year's exposure, are, in most instances, not so great as to counterbalance the profit arising out of laying in stocks at a convenient time.

CHAPTER XV.

Description of the Snow Shoe district of the First Bituminous Coal Basin in Centre County, Pa.

The Eastern sub-division of the first coal basin has been described in its development in the Osceola and Philipsburg region. From Philipsburg the basin extends north-east, the Lower productive coal measures occupying only a narrow belt, and the underlying non-productive coal measures rising high on the hills which can only take in on their extreme tops the lowest beds of coal. This narrow line of coal measures continues to the Snow Shoe Coal Basin, where the enclosing conglomerate rock widens out, and the basin deepens and takes in over a very considerable area the same measures already described as showing near Osceola.

North-eastward of Snow Shoe, these coal measures continue to occupy the hills or hill tops through the Beach creek, Tangascootack and Queen's run regions. They are all in Mr. Hodge's First Bituminous Coal Basin, having the Allegheny mountain axis for the south-east line of their conglomerate bottom rock and the First Anticlinal sub-axis as the limit to the north-west.

It has already been stated, that the First Coal Basin is subdivided into two distinct sub-basins by an anticlinal sub-axis which runs along its centre, parallel to its two sides. This sub-axis passes between Snow Shoe and Karthaus, making the Snow Shoe Basin the eastern sub-division of the First Basin, while the Karthaus Coal Basin is the western sub-division. Considered as a whole at this point the Allegheny mountain forms the eastern side of the First Basin, and the dying First axis, or Laurel Hill axis, 5 to 7 miles west of Karthaus, forms the western side.

The accompanying map of the Snow Shoe Basin, (Plate VI,) is for the most part compiled from original work done by Mr.

Jas. L. Somerville, Chief Engineer of the Bellefonte and Snow Shoe railroad. The map with the accompanying cross section, will serve to explain the report; and the cross section map, (Plate II,) from Snow Shoe to the Nittany mountain, illustrates the structure of the eastern escarpment of the Allegheny mountain at this point.

The Seral conglomerate of Rogers, (No. XII,) the base rock of the Lower Productive Coal Measures, forms the crest of the Allegheny mountain at the Snow Shoe, and is made up at that point, as at Karthaus, of about 250 feet of massive sandstones, with some pebble rock layers of rounded white quartz pebbles of various sizes, from a pea to an egg. These conglomerate rocks dip very slowly to the north-west, in fact are almost horizontal, and no coal outcrop of any consequence is found for several miles north-west of the mountain crest, the nearest one being some three or four miles from it.

One and a half miles south-east of Snow Shoe village, the bottom Conglomerate shows on the surface in lumps and boulders, and the basin is sharply edged up in that direction. To the north-east of Snow Shoe, the map shows the Lower Productive Coal Measures continued in Sugar Camp Hill, and two miles still further to the north-east the shales and thin sandstones exposed take in Bed A, and perhaps Beds A and B of the same measures. At the Beech Creek crossing, massive conglomerate boulders show in the stream bed. On the north of the basin, the bottom Conglomerate shows east of Germania, cutting off the productive measures to the north and north-west, while the Little Moshannon creek may be roughly taken as the limit to the south and south-west. For the basin *rises in nearly all directions* from its centre at or about the Snow Shoe mines, and though a small black slate show is found high on the hill crest south of the Little Moshannon, there can only be there the lowest of the productive coal measures. The basin, therefore, as already stated, terminates to the north-east in the narrowing and interrupted basins leading towards Farrandsville, and to the south-west in the narrow and shallow, and in most cases worthless line of coal measures which connect it with the Philipsburg and Osceola coals.

The centre of the basin is somewhere near Lucas and Askey

Hills, and it is at that point, therefore, that the deepest exposure of measures is found.

The section as determined by Mr. Sommerville, by shafting, is as follows:

Fig. 31.

Snow Shoe.

Surface and cover,	25	
Coal, E. Upper Freeport,	5	
Fire-clay,	5	
Concealed measures,	41	9″
Sandrock,	7	
Coal, Middle Freeport,	2	
Brown ore and coal,	2	
Hard limestone,	2	6
Light colored clay,	5	
Black slate,	34	4
Slaty cannel,	2	
Coal, D, Lower Freeport,	5	8
Fire-clay,	3	
Slates, light colored,	16	6
Light colored sandstone,	13	
Coal, slate, "slaty coal," C,	4	
Black slate,	11	6
Grey sandstones,	17	
Black slate,	10	
Coal, bony,	1	
Coal, B,	4	
Concealed measures,	45	
Iron ore,	4	
Grey sandy slates,	14	
Coal, A,	3	

Of the coal beds in the above section, the opening on Bed A is now fallen shut and the coal cannot be measured. It is reported to have yielded 3 feet to 3½ feet of good coal: and to have been quite free from sulphur.

Coal Bed B of the section (Fig. 32) is opened and worked near the north-east point of Coal Hill. It is called locally the "Lower Bed."

Fig. 32.

Lower Snow Shoe.

A measurement in the mines gives:—

Black slate roof, hard and good.		
Bony coal,	0′	11″
Coal,	2	3½
Slate, not persistent,	0	1
Coal,	0	6
Slate, persistent,	0	1
Coal, sulphurous,	0	9
Fire-clay,	1	0
Coal, left in floor,	10″ to 12	
Fire-clay floor.		

The specimen forwarded for analysis yielded, (M'Creath):

"Water,	0.750
Volatile matter,	23.440
Fixed carbon,	64.374
Sulphur,	.986
Ash,	10.450
	100.000

Coke, per cent., 76.80. Color of Ash, grey, with red tinge.

The coal is bright, shining, contains veins of charcoal and Pyrites. Iridescent."

The specimen forwarded for analysis could not have represented a fair average of the mine yield: the percentage of ash being entirely too high.

The coal is chiefly used for steam raising, being shipped almost exclusively for that purpose.

Coal Bed C of the section is nowhere opened up in the basin. Where struck in shafts and trial pits it has always been found slaty and worthless, usually showing as a mass of slate and coal of 4 feet or more in thickness. The development of the black slate between coal beds B and C is a noteworthy feature.

Coal Bed D of the section, known locally as the "middle bed," or "big bed," has been opened and extensively worked.

Fig. 33.

Middle Snow Shoe.

Where measured in the mine it shows:

Black slate roof.	
Slaty cannel coal, worthless,	8″ to 2′
Coal,	2 4″
State, persistent,	6
Coal,	1 6
Slate, persistent,	1
Coal,	6
Fire-clay floor.	

The upper bench of the coal is harder, and breaks out somewhat in blocks; the lower benches are friable, columnar in structure, and seem rather purer.

An average specimen yielded, on analysis, (M'Creath):

"Water,	0.650
Volatile matter,	24.560
Fixed carbon,	70.416
Sulphur,	.964

Ash, - - - - - - - -	3.410
	100.000

Coke per cent., 74.79. Color of Ash, cream.

The coal has a resinous lustre, contains thin veins of Pyrites and Charcoal."

The coal slack is roughly coked in the open air near the mine mouth, and yields a firm, compact coke. But it is carelessly done, and much slate gets in with the coal slack.

A specimen of this coke yielded on analysis, (M'Creath):

"Water, - - - - - - - -	.990
Volatile matter, - - - - -	2.950
Fixed carbon, - - - - - -	82.626
Sulphur, - - - - - - -	1.104
Ash, - - - - - - -	12.330
	100.000

Color of Ash, red.

The coke has a dull grey lustre, is very hard and compact, and with much slate in small pieces."

The small coal lying directly on top of the Freeport Limestone (the Middle Freeport coal of the section) is at this place, as ordinarily elsewhere, an entirely worthless bed.

The Coal Bed E of the section, (Upper Freeport,) the "upper bed" locally, is opened and worked at Mine No. 5. Where measured in the mine it showed:

Fig. 34.

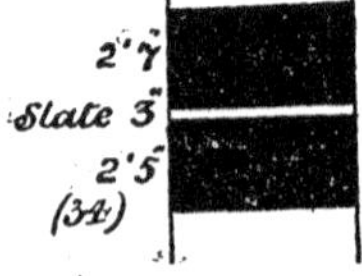

Roof slate.		
Coal, - - - - -	2	7"
Slate parting, persistent, -	2'	to 4
Coal, - - - - - - -	2	5
Fire-clay floor, very soft.		

An average specimen of this coal yielded on analysis (M'-Creath):

"Water, - - - - - - -	1.280
Volatile matter, - - - - -	25.580
Fixed carbon, - - - - - -	68.937
Sulphur, - - - - - - -	.613
Ash, - - - - - - -	3.590
	100.000

Coke per cent., 73.14. Color of Ash, cream.

The coal is compact, with a shining lustre, very hard, and containing small veins of charcoal."

The great thickness of the black slate overlying the Middle bed (it is 34½ feet thick) is remarkable.

The coal of the Upper bed is said to have made excellent coke; and is especially shipped for use in working iron

It may be noted of these three coal beds, B, D and E, as worked in Coal Hill (B and D) and in Lucas Hill, (B, D and E,) that the upper bed is very decidedly troubled; both floor and roof being irregular and uneasy and pinching down. In fact these irregularities interfere somewhat with profitable working.

These troubles are not met with on the Middle bed, the only difficulty showing, so far as worked, being a heavy and troublesome roll of 15 feet at Mines 6 and 7. When this was overcome everything was found smooth and regular as before. Mr. Sommerville states that in this case the trouble in the Upper bed corresponds with the trouble in the Middle bed, and that in fact it is a regular downthrow fault of 10 feet, having a general course of north 8° east, and south 8° west.

Bed B has not been very extensively worked as yet, but appears not to be troubled with irregularities.

The "Middle Bed," Bed D, was opened at Gonzales Mine at the time of the First Geological Survey of the State, and is mentioned in the Rogers' Final Report as showing 6 feet of coal, including a layer of cannel coal, 6 inches thick, running through the middle; with a roof of slaty cannel and the body of the bed made up of a beautifully brilliant and pure glance coal, dipping, strange to say, to the north-east. Such an abnormal dip could only have been very local.

This Gonzales mine is now fallen shut. It was opened on the "Middle Bed," the Lower Freeport Bed.

The measures as given in the section cross over the ravine of Askey's Run and come into Askey's Hill, the upper bed lying close to the hill top: the rise to the north and north-west soon carrying it out into the air.

The Snow Shoe Mining Company are now putting trial shafts on the east end of Askey's Hill. The slight opening made indicates that the coal measures there still dip slightly to the north-west, making Askey Hill about the centre of the synclinal, for all the coals opened on the north-west side of the hill dip back decidedly to the south-east.

The coal basin rising, as has been stated, in all directions from its centre, the lower coals of the Lower Productive Coal Measures all come to daylight north of the Little Moshannon creek. The openings on the large bed (the Lower Freeport) on Snow Shoe Coal Hill have hitherto supplied entirely the small local demand of the region, and no effort therefore has been made to open up and work the coal beds in the vicinity of Moshannon village. The outcrops show and some few openings were made many years ago: but they are all now fallen shut.

One mile north of the village a shaft on the north-west side of Askey's Hill, back of Holt's house, opened up a 30 inch coal and another small bed outcrops in the road near his house, the vertical distance between the beds being about 40 or 45 feet.

Hoy's mine is on the east side of Seven Mile Run, west of Holt's house. The mine ran in to the south-east, which is down the dip, and is now drowned out. It is reported as a "3 foot bed." On the west side of Seven Mile Run is a broad, high hill, on the west side of which the coal beds begin to come to daylight finally, making the west or north-west side of basin.

Two miles north-east of Moshannon village, Wm. Holt has opened up a small coal.

The measures show imperfectly at his place and give the following section, (Fig. 35):—

Fig. 35.

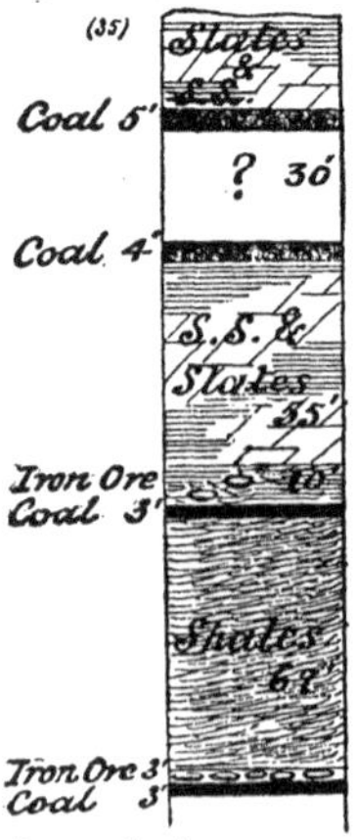

Hill top.	
Drab slates and thin sandstones,	15′
Coal,	"5," not seen
Concealed measures,	30
Coal,	"4," not seen.
Thin slaty sandstones and gray slates,	55
Black slates and shales, with lean hematite,	10
Coal,	"3," not seen.
Brown and gray shales and *concealed measures,*	67
Iron ore balls,	"3"
Coal,	3
Water level of creek.	

Of these coal beds only the lower one is open and worked. The size of the beds above is given only as reported by Mr. Holt, none of them being seen.

In the opening on this lower bed (Fig. 36) the coal measured:—

Fig. 36.

Black slate roof, - -	1′ 6″
Coal, - - - -	6″ to 8
Slate, - - - - -	8½
Coal, - - - -	2 4
Fire-clay, - - -	2 6

The coal averages from 28 to 30 inches in thickness in one bench, the upper 16 inches being of columnar, friable coal and the lower 12 to 14 inches, very hard. As this was the only opened mine in the north-west part of the basin specimens were taken for analyses from the upper and lower parts of this bench.

The upper bench yields, (M'Creath):

"Water, - - - - - -	1.680
Volatile matter, - - - - -	21.870
Fixed carbon, - - - - -	71.108
Sulphur, - - - - - -	.612
Ash, - - - - - - -	4.730
	100.000

Coke per cent., 76.45. Color of Ash, red.

The coal has a shining lustre generally, some pieces dull metallic lustre, iridescent."

The lower bench yielded, (M'Creath):

"Water, - - - - - -	0.880
Volatile matter, - - - - -	23.620
Fixed carbon, - - - - -	70.089
Sulphur, - - - - - -	.661
Ash, - - - - - - -	4.750
	100.000

Coke per cent., 75.50. Color of Ash, red.

The coal has a very dull lustre, and shows considerable iridescence."

The above analyses show an excellent character of coal.

The next coal bed, 70 feet above, called a "3 foot bed," is not now open, but good, bright, firm bituminous coal from it was found lying alongside of the trial shaft.

The opening on the next coal above, 135 feet above the run,

is fallen entirely shut. It is reported a "4 foot bed," and "much troubled by clay horsebacks and by pinching."

Nothing shows at the old abandoned mine on the next bed above, 165 feet above the run. It is reported as a "5 foot bed," called locally the "6 foot bed."

East of Germania large masses of Conglomerate show in the road, marking that the Lower productive coal measures have risen into the air here and passed out of the hill tops.

On the Martin Long place, on the west side of the basin, a "3 foot" bed of coal, opened and worked many years ago, is now fallen shut. It is apparently the same as the Hoy mine, already described. In the bottom below this old opening is a lean bog ore deposit. It is not extensive, and the ore is very light. Where the road crosses Seven Mile run, (at the junction of Snow Shoe and Burnside townships) a "3 foot" bed was opened in the creek bank many years ago.

The five foot bed outcrops back of the School House in the hollow between Lucas and Askey Hills.

Going north-west along the Snow Shoe and Karthaus pike, numerous outcrops of black slate and smut show themselves on the roadside. These are the lower beds of the Snow Shoe section, and are gradually rising to the north-west. These coal crops are seen north-west of J. Rapp's house, and near Beightol's house; and an iron ore show is found near the cross-roads, at Stewart's house. But the sandstone grows more massive on the roadside, and at or about the old Pine Glen the First Anticlinal sub-axis is crossed, massive fine grained Conglomerate making the surface rock everywhere around.

Crossing the North Fork of Beech creek, at the north-east end of Coal Hill, we find the measures on Sugar Camp Hill partially opened up, only just sufficiently to show that the coals of the Snow Shoe section, (except the upper one,) exist in the hill as workable deposits.

The general nature of the coal deposit in the Snow Shoe region, the limited amount of the upper Freeport Coal Bed, the much greater amount of the Lower Freeport Coal Bed, and the very broad extent of the lower beds, A and B, are shown in the cross section and dotted outcrop lines of the sketch map of the basin, (Plate VI.) These dotted lines are only approximately correct, the main features, however, being kept sufficiently close to fact, by those numerous mines and trial shafts which are accurately located.

Snow Shoe Iron Ores.

The vertical section of the Snow Shoe measures, (Fig. 31,) shows the Freeport iron ore resting on top of the Freeport limestone; the black band iron ore in the shales overlying coal bed C, and the iron ore overlying coal bed A.

The Freeport iron ore as found in the shaft and seen upon the outcrops averages about two feet (in all) of "ore and some coal," the Middle Freeport coal two feet thick, resting directly on top of it. A specimen of the ore from Yeager's place, west of Snow Shoe City, on analysis at the laboratory of the survey, yielded (M'Creath):

Iron, - - - - - - -	30.250
Sulphur - - - - - -	.112
Phosphorus, - - - - - - -	.211
Insoluble residue, - - - - -	19.630

A specimen forwarded by Mr. Sommerville, was probably taken from the underlying limestone, as it yielded (M'Creath):

Iron, - - - - - - -	5.000
Sulphur, - - - - - -	.599
Phosphorus, - - - - - - -	.050
Carbonate of lime, - - - - -	42.941
Carbonate of magnesia - - - -	22.764
Insoluble residue, - - - - -	18.730

A specimen of iron ore from M'Master's place, near Snow Shoe City, yielded (M'Creath):

Iron, - - - - - - -	35.800
Sulphur, - - - - - -	Trace.
Phosphorus, - - - - - - -	.204
Insoluble residue, - - - - -	16.050

A fair specimen of the Freeport limestone of the Snow Shoe Basin was forwarded to Mr. M'Creath for analysis, and yielded:

Carbonate of lime, - - - -	51.153
Carbonate of magnesia, - - - -	13.265
Sulphur, - - - - - -	Trace.
Phosphorus, - . - - - - -	.287

The iron ore underlying the Lower Freeport coal, D, or rather overlying Bed C, is a black band iron ore, of good quality. It is called a "20 inch to 36 inch" ore bed, as found in the shaft. Mr. M'Creath did not determine the percentage of carbon carried by the ore, but reports the other constituents thus:

Iron,	29.300
Sulphur,	.010
Phosphorus,	.201
Insoluble residue,	17.60

When roasted, this ore yielded in the laboratory of the survey, 43 *per cent. of metallic iron.*

Mr. Sommerville also forwarded to the laboratory a specimen of this ore, (labeled clay band ore,) and it yielded on analysis, (M'Creath:)

Iron,	28.700
Sulphur,	.011
Phosphorous,	.178
Carbonate of lime,	——
Carbonate of magnesia,	——
Insoluble residue,	23.600

The iron ore overlying Coal Bed A is not now worked.* From the old opening it was possible to procure specimens, but not to verify the reported thickness of the ore bed, "4 feet in all."

Two specimens yielded on analysis, (M'Creath):

	No. 1.	No. 2.
Iron,	30.100	32.600
Sulphur,	.086	.013
Phosphorus,	.364	.993
Carbonate of lime,	——	——
Carbonate of magnesia,	——	——
Insoluble residue,	23.250	20.530

*This bed was opened up some years ago, and in a report to the Snow Shoe Mining Company, by Morgan C. Davis, was described as follows:

"Assay of the argillaceous iron stone found in shaft below Dutch Hen's coal opening, (now known as Mine No. 4,) on the road to the New Saw Mill. This iron stone loses by calcination, 23 per cent.; 400 grains of the calcined iron stone yielded 168 grains of iron, or 42 per cent.

"The iron stone is found to be 4 feet 6 inches thick in the trial shaft in which it was found. It is irregular there on account of a fault (*sic*) coming in near the centre of the shaft, which presses the iron stone down to less than one-half its thickness."

Just west of the summit of the Allegheny mountain, and in the bottom of the Seral Conglomerate of XII, pieces of hematite iron ore cover the surface, and are found a little below it, over a limited area. The line of ore is not found extending along the strike of the measures, and there is no indication that a regular and persistent workable deposit is to be looked for.

A specimen of the ore yielded on analysis, (M'Creath):

Iron,	41.000
Sulphur,	trace.
Phosphorus,	.692
Insoluble residue,	25.250

The above descriptions and analyses indicate a very favorable outlook for the iron ores of the Snow Shoe Basin. They are in considerable quantity, and from the openings at different points apparently fairly regular in thickness; and their percentages of metallic iron and low percentages of phosphorus and sulphur suffice to make them of excellent quality.

For with the carbonate iron ores, and especially the "black band" iron ore in position to be supplied cheaply to furnaces in the Bald Eagle valley; with the rich hematite iron ores of the Lower Silurian limestone valley, east of Bellefonte, in abundant supply; and with the Snow Shoe coals at hand to make a cheap and good coke, it seems clear that the time for the use of the Snow Shoe iron ores, in large quantities, cannot be far off.

Snow Shoe Coals.

The coal from the Snow Shoe basin is so similar in character to that shipped from the First Coal Basin at Houtzdale, Osceola and Philipsburg, that it is only necessary to refer to Chapter XIV of this report, in which the nature of the steam coal from this First Basin is briefly considered.

The Snow Shoe region ships about 100,000 tons of coal to market annually; used mainly for steam generating purposes and in iron works; largely in the Lehigh valley; and for locomotive use on the Philadelphia and Erie railroad, where it gives entire satisfaction as a strong burning steam coal.

Appended is a table of the analyses of coals and cokes from the Snow Shoe Basin, arranged for convenient comparison with the tables given in Chapter XIV.

Snow Shoe Coals.

No.	Water.	Volatile Matter.	Fixed Carbon.	Sulphur.	Ash.	Color of Ash.	Coke, Per cent.
1......	.88	23.62	70.089	.661	4.75	Red	75.50
2......	1.68	21.87	71.108	.612	4.73	Red	76.45
3......	1.28	25.58	68.937	.613	3.59	Cream	73.14
4......	.65	24.56	70.416	.964	3.41	Cream	74.79
5......	.75	23.44	64.374	.986	10.45	Gray, with red tinge,	76.80

No. 1. Wm. Holt's mine, west of Holt's hill, 2 miles north-west of Snow Shoe City. Bottom bench of bed, lower hard part of bench.

No. 2. Wm. Holt's mine, Snow Shoe Basin, 2 miles north-west of Snow Shoe City. Upper part of bench.

No. 3. Snow Shoe, mine No. 5, Upper Bed.

No. 4. Snow Shoe, mine No. 6, Middle Bed.

No. 5. Snow Shoe, mine No. 4, Lower Bed.

Snow Shoe Cokes.

No.	Water.	Volatile Matter.	Fixed Carbon.	Sulphur.	Ash.	Color of Ash.
1................................	.990	2.950	82.626	1.104	12.330	Red.

No. 1. Snow Shoe R. R. Co.'s Colliery, mine No. 6, Middle Bed, coked in open air from coal slack.

CHAPTER XVI.

Description of the Karthaus district of the First Bituminous Coal Basin in Clearfield county.

The First Anticlinal sub-axis, carrying the massive sandstones of the Seral Conglomerate or Millstone Grit high upon its crest, has its centre near the old "Pine Glen," on the Snow Shoe and Karthaus turnpike. This axis divides the First Coal Basin as the Viaduct anticlinal axis divides the same basin in Cambria county; the Snow Shoe coals representing the eastern sub-basin in Centre county, as the Bennington, Lilly's and Ben's creek coals do in Cambria county; while the Karthaus coals represent the western sub-division of the First Basin in Clearfield county, as the Johnstown coals do in Cambria county. Hitherto Karthaus has been incorrectly placed in the Second Coal Basin.

Going north-west from "Pine Glen" to Karthaus, the lowest coals in and above the seral conglomerate are found at new Pine Glen (Boak's) in a well; at George Boak's, also in a well, and with black slate outcrops on the roadside just beyond; on the roadside beyond Mulholland's house and near Loy's house. All these are dipping decidedly to the north-west; and by the time the West Branch of the Susquehanna is struck, the massive conglomerate has come down from the hill tops and makes the bottom and sides of the river. It is here made up of coarse massive sandstone, with only occasional pebbly layers.

At the time of the First Geological Survey of Pennsylvania the measures at Karthaus were very completely opened up by the Karthaus Iron Company, and were thoroughly examined. The furnace has been out of blast for many years, the mines are fallen shut, and it is not now possible to procure so complete a vertical section of the measures as the one made at that time. It is, therefore, reproduced here from the Pennsylvania Final Report, as follows:

Fig. 37.
Karthaus.

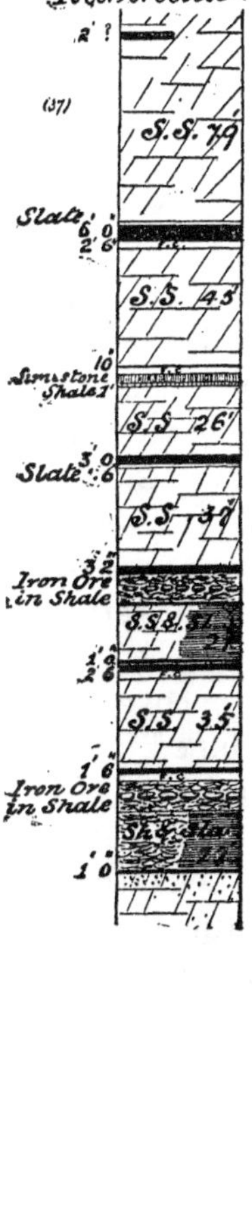

Slaty sandstone (up to the summit of the hill, 565 feet above the river) said to contain a coal bed 2′ thick,	79′
Blackslate,	1
Coal, (elevation 479′)	6
Fire-clay, poor,	2 6″
Brown sandstone,	45
Coal,	0 10
Fire-clay,	2
Limestone, silicious,	3 6
Shale,	1
Brown sandstone,	26
Coal,	3
Slate,	1 6
Gray sandstone,	37
Coal,	3 2
Shale, containing 26″ of good kidney form iron ore (elevation 345′)	11
Coal,	1
Brown sandstone and slate,	21
Coal,	1
Slate,	0 3
Coal,	2 6
Fire-clay,	2 6
Brown sandstone,	35
Coal,	1 6
Fire-clay, ferruginous,	3
Shales, containing 25″ of good iron ore, called the "Red ore band" (elevation 268′)	11 9
Shale and slates,	22
Coal,	1
Sandstone, the "brown rock."	
Coal, thin.	
Seral conglomerate down to the river, 240′.	

Fig. 38. Reiturs' Bank.

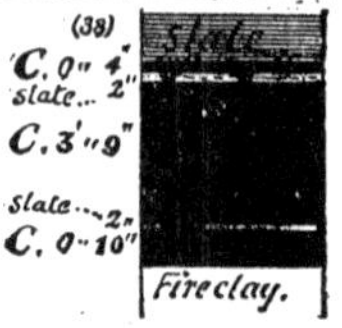

The only coal now opened at Karthaus is the "6 foot bed," about 500 feet above the river. Where worked at Reitur's Mine it shows:

Roof, black slate.	
Coal, - - - - - - -	0' 4"
Slate, - - - - - - - -	0 2
Coal, - - - - - - - -	3 9
Slate, - - - - - - -	1" to 2
Fire-clay floor.	

The mine showed an irregular and uneasy floor; the fire-clay coming up in one place, and for a distance of 30 yards cutting the coal down to 2 feet. But it is said to be usually a quiet and regular coal bed.

An average specimen yields on analysis, (M'Creath:)

"Water, - - - - - - -	.630
Volatile matter, - - - - -	24.630
Fixed carbon, - - - - - -	70.396
Sulphur, - - - - - - -	.654
Ash, - - - - - - - -	3.690
	100.000

Coke per cent., 74.74. Color of Ash, red.

The coal is bright, shining, very hard, with small scales of Iron Pyrites."

The above analysis is very flattering and fully confirms the statement in the Pennsylvania Final Report, that this "6 foot Bed" was adapted to making a very superior coke. Besides this upper coal, there are three others of a size suitable for mining, one of which is 3½ feet thick, and the two others each 3 feet; but none of these supply as pure a coal as the main seam.

The limestone in this neighborhood is for the most part inferior, and only one bed of it, 3½ feet thick, occurs in the series.

Lower down in the Measures occur two important beds of iron ore. One of these, at an elevation of 345 feet above the river, is estimated to contain in all, 2 feet of good blue kidney ore in 11 feet of shales. The other lies at an elevation of 268 feet above the river. It also exhibits about 2 feet of good kidney ore in a stratum of shale less than 12 feet thick. This band is locally called the "Red Ore," and is of a different variety from that above it; they are both of excellent quality.

This ore, (mottled brown, nodular concentric, crust hematitic,) gives by analysis:

Carbonate of Iron, - - - - -	19.86
Peroxide of Iron, - - - - - -	34.80
Carbonate of Lime, - - - - -	4.50
Silica and Insoluble matter, - - -	30.40
Alumina, - - - - - - -	1.70
Water, - - - - - - -	8.20
Metallic Iron in 100 parts, - - -	33.95

The above description is taken mainly from the Pennsylvania Final Report.

"A specimen of this "red ore," forwarded to Mr. M'Creath for analysis, yielded:—

Insoluble residue, - - - - -	21.040
Iron, - - - - - - - -	34.000
Sulphur, - - - - - - -	.054
Phosphorus, - - - - - -	521

The ore is a carbonate, partially converted into Oxide on outside."

Prof. Walter R. Johnson made analyses (in 1838) of the minerals found at Karthaus. He reports as follows:—

"Six foot Coal Bed."

Specific gravity, - - - - - -	1.250 to 1.278
Loss of water in distillation, - - - -	0.60
Carburetted Hydrogen and other volatile products,	26.20
Earthy residuum after complete incineration, -	5.05
Carbon in the coke, - - - - - -	68.15
	100.00

The coke is of medium hardness and in all respects well adapted to the production of iron. The earthy residuum is composed of silex, alumina and oxide of iron with a portion of lime and a little magnesia."

In another analysis made in 1844, when testing the efficiency of American coals in the generation of steam, for the United States Navy Department, Prof. Johnson gives for the Karthaus "6 foot Bed:"

Specific gravity, - - - - - -	1.284
Volatile combustible matter, - -	19.530

Fixed carbon, - - - - - -	73.770
Earthy matter, - - - - - -	7.000

An analysis of this coal, made for the First Geological Survey of Pennsylvania, is published in Rogers' Final report as follows:—

Volatile matters, - - - - - -	24.800
Coke, - - - - - - - -	75.200
Ash, - - - - - - - -	4.700

Prof. Johnson's analysis of the Kidney ore is—4 analyses:—

	1	2	3	4
Specific gravity, - - - - - - - - -	3.397	3.415	3.206	——
Of water at 320° - - - - - - - - -	1.200	3.900	27.420	27.42
Carbonic acid by calcination at red heat, -	26.680	6.720		
Metallic iron, - - - - - - - - -	38.330	50.600	36.100	34.54
Earthy impurities, silica, &c., - - - -	16.670	17.100	26.170	27.34
Specific gravity of pig metal obtained, - -	7.726	6.240	7.102	——
Oxygen, - - - - - - - - - - -	——	21.680	10.310	——

Specimen No. 2, was the shell of the carbonate of iron weathered on the outcrop to a brown hydrate of the peroxide of iron.

His analysis of the "Red Vein" is:—

	1	2
Specific gravity, - - - - - - - - - - - - -	3.421	3.421
Loss by calcination, water, carbonic acid, &c., - - - -	29.060	29.060
Metallic iron, - - - - - - - - - - - - -	35.910	36.070
Earthy impurities, silex, alumina, &c., - - - - - -	20.680	20 380
Oxygen and other volatile products of fusion, - - -	14.350	14.490
Specific gravity of pig metal, - - - - - - - - -	6.787	7.272

Prof. Johnson's analysis of the Upper Limestone (Freeport Limestone) is as follows:—

Specific gravity, - - - - - -	2.780
Loss by calcination, water and carbonic acid,	36.370
Dry Lime contained, - - - -	36.080
Protoxide of iron, - - - -	6.970
Silica, - - - - - - - -	12.000
Alumina and Manganese, - - -	8.580

The abundance of these materials, their excellent character, and their immediate proximity to each other, all in the same hillside, naturally point out this Karthaus region as a place for the manufacture of iron. This is especially worthy of note now, when the project of building a railroad up the West

Branch of the Susquehanna river from Keating, on the Philadelphia and Erie railroad, is being strongly urged.

The basin extends north-east from Karthaus, down the Susquehanna, for several miles; the large upper bed of Karthaus entering the hills above the neighborhood of Three Runs. At the latter point a bed of coal 3 feet 2 inches thick has been opened, associated with a layer of lime and one of fire-clay. Just north-east of Karthaus Mr. Heckendorn works a "5 foot" bed; and a coal mine (now fallen shut) was worked at Snar's mill. A boring made on the West Branch of the Susquehanna, 2 miles below Karthaus, is thus reported:—

Surface.		
Sandstone - - - - - - - -	72′	
Coal, - - - - - - - - - - -	1	10″
Sandstone, white and red, with soft rocks at bottom, - - - - - - - - -	232	
Grey limestone rock.		
Strong salt-water at 393′ below the surface,	393	

On Birch Island Run, which enters the Susquehanna below Karthaus, near the county line, several beds of coal were developed many years ago. The most important of these was opened on the hill between the forks of the stream. The bed is reported to show 6 feet of coal, and is regarded as the big Karthaus bed. A still higher seam, 4 feet thick, is found on the hill top, about 30 feet over the large bed. Iron ore is found associated with the shales enclosing it. Beneath the large coal, 40 feet, a stratum of limestone, 3 feet thick, outcrops. On the same property farther west, two other coal beds have been opened. They are reported as measuring 2½ and 4 feet respectively, and occupy a position lower in the measures than the large seam. These openings are all long since fallen shut, and are re-described from Rogers' Final Report.

Nine miles above Karthaus, on the river, the "red ore" has been found, and is stated to be thicker than at Karthaus. Three miles higher up there occurs a bed of coal near the summit of the highest hills, which measures 4 feet 4 inches in thickness; this, there is some reason to suppose, is part of the main Karthaus seam. Between the two localities mentioned appears a bed of limestone. In this neighborhood the seral conglomerate and sandstone occupy a position near the flats of the river.

The coal measures (Lower Productive) spread but a moderate distance from the river on either side, the underlying conglomerate rising out and capping the hills everywhere from 3 to 5 miles from the river. (Hodge in Rogers.)

The First Anticlinal sub-axis running north-east through Morris township, crosses the Moshannon not far above its mouth. Crossing the high land of this axis between the West Branch of the Susquehanna and the Moshannon creek, the extreme north point of the lower productive coal measures of the First Basin is found somewhere near the head waters of Crawford run. For though the First Basin continues on to the Snow Shoe, yet between the above point and the Little Moshannon the conglomerate is near the hill tops, and the productive basin narrow and shallow.

Near this point (at headwaters of Crawford run) a bed of coal was opened many years ago, but is now fallen shut. The blacksmith reports it as working well in his fire, and as being "6 feet thick." There is no way of judging surely, but this should be bed A, of the First Basin section.

Going south-west from this point the basin deepens, taking in higher coal beds. At Keffer's mine, ½ mile west of Kylertown, the coal shows:—

Fig. 39.
Keffers Bank

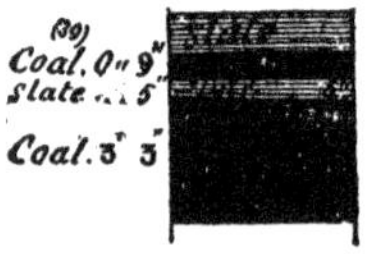

Black slate roof, good.

Coal, - - - - -	9′.
Slate, - - - -	4 to 6″
Coal, good, - - -	3′ 3

Fire-clay floor.

The coal looks and mines out well. The measures are dipping gently to the south-east; the mine being on the north-west side of the First Basin. The bed looks much like coal bed B of the First Basin section.

Mr. J. Potter opened up a coal bed on his place, ½ mile north-east of Kylertown; and coal has also been opened on A. Brown's place, on the head waters of Grass Flat run. Two workable beds are reported at the latter place, separated by only 20 feet of measures.

Limestone was once opened at Kylertown, (only a few feet above the road,) but it contained too much iron to burn well for agricultural purposes. Over the limestone is a small coal smut. None of the sections made in the First Basin to the south-west of this place have shown this limestone bed.

At Mons' bank, ½ mile south of Kylertown, two coal beds show, separated by thirty-five feet of measures. The lower bed shows:—

Fig. 40.

Mons' Bank.

Black slate roof,	5′	
Cannel slate,		8″
Coal,	1	2
Cannel slate,	1	3
Coal,		6
Fire-clay floor, hard.		

The coal is hard and has a cannel look, but is in no respect a cannel coal. An average specimen analysed (M'Creath):

"Water,	0.750
Volatile matter,	19.570
Fixed carbon,	69.833
Sulphur,	.677
Ash.	9.170
	100.000

Coke per cent., 79.68. Color of Ash, gray with reddish tinge.

The coal is dull, with resinous lustre, very compact and hard."

Where this coal is opened again, some 200 yards away, the cannel appearance has entirely disappeared, and the coal shows simply as an ordinary and rather soft bituminous coal.

The bed opened 35 feet above, shows only 18 inches of coal; 80 feet above the upper of these beds is a marked bench, which is apparently the bench of the limestone bed.

Between Kylertown and Morrisdale, the outcrops of these small beds show occasionally on the roadside; and also west of Morrisdale, until upon reaching the high land just east of Centre Hill, the surface of the ground is covered with conglomerate lumps, and the centre of the First Anticlinal sub-axis is found.

West of Centre Hill outcrops of black slate, and occasionally an insignificant coal smut show at Centre Hill, at Smael's, and at Williams' Grove, (Bigler station on the Tyrone and Clearfield railroad.) At the latter place the hills rise from 200 to 250 feet above the creek. Two benches show plainly, and it is reported that a 3 foot coal bed was opened on one of them.

On the south-eastern side of this western sub-division of the First coal basin the measures are almost flat, with an average very gentle dip to the north-west. But there seem to be several slight rolls in these almost horizontal measures which about neutralize the sinking to the north-west and keep the same measures on the level, pitching softly first in one direction and then in the opposite one.

Passing west and north-west along the Tyrone and Clearfield railroad from Blue Ball station, the point where the First Anticlinal sub-axis crosses the railroad, about ½ mile beyond the station, is marked by a beautiful exhibition of the seral conglomerate or millstone grit. Enormous boulders of fine grained white quartzose sandstone, with some brownish massive sandstone, are found, and occasional massive layers of conglomerate rock with rounded white quartz pebbles of the size of a pea and larger. The mass rises as a wall 50 to 60 feet high. Some of the loose blocks will contain over 2,000 cubic feet. As exposed here, this mass of sandstone and conglomerate should be in all some 200 or more feet in thickness.

The railroad, following the stream, keeps in this conglomerate, sometimes dipping softly in one direction and then back again, or about flat until near Wallaceton; where overlying measures come in, and coal is found outcropping. In wells in the village a small coal is struck only a few feet below the surface, with from 6 to 12 feet of fire-clay underlying it. Where the lowest exposed coal was struck in a well, about 500 yards south-west of Wallaceton, it shows about 2 to 2½ feet of coal with fire-clay floor and sandy gray slates for cover. Jacob Smael's mine had fallen shut; but the props of the main entry were only 2 feet 9 inches high, and the bed of course must have been small. The dip, at this point, is slightly back to the south-east.

At Shimmel's opening, two-thirds of a mile north-east of the station, the main entry has fallen in; but from the size of the opening the bed could not have been large. Gray slates overlie the bed. On the hill south of this mine, two small beds were once opened up, dipping to the south-east gently.

At Ross Summit, half a mile north of Wallaceton, a railroad cut shows:—

Top of cut.		
Thin bedded sandstone, - -	8′	
Sandstone, - - - - - -	1	
Impure fire-clay, - - -	4	
Black slate, - - - -	3	6″
Coal, - - - - - - -	0	6
Fire-clay in bottom, - - -	4	

This cut shows the unevenness of the measures in this neighborhood; the summit being a small synclinal, the rocks on either side dipping gently in to the centre of the cut.

Thence westwardly to Clearfield town about the same measures are exposed, the conglomerate, however, being below the river at Clearfield town, but showing some small pieces on Clearfield creek.

CHAPTER XVII.

Description of the Second Bituminous Coal Basin of Clearfield county, around Clearfield and Curwensville.

The coal beds at and near the town of Clearfield have been considerably developed for local use, though no coal is shipped by railroad to market.

The section (Fig. 41) shows the measures exposed on Shaw's Hill, east of the extreme north end of the town, as follows:

Fig. 41.

Shaws Hill.

Hill top.		
Sandstone, - - - - - -	20′	
Shales, (with impure limestone?)	16	
Coal, - - - - - - - -	2	
Shales, - - - - - - -	51	
Coal, - - - - - - - -	3	4″
Fire-clay and shales, - - -	6	
Fire-clay, in all, - - - -	24	
Sandstone, - - - - - -	5	
Black slate, - - - - - -	7	
Coal, Peacock, - - - - - -	0	7
Concealed measures, . - -	14	
Susquehanna Rivel level.		

The main coal bed of the section measured as follows:

Shales.		
Roof, black slate.		
Coal, - - - - -	1′	5″
Slate, - - - - -	0	1½
Coal, - - - -	1	11½
Fire-clay (?) floor.		

The parting slate layer thickens in places to 2 and 3 inches, and runs down at times to 1 inch. The coal is bright and shining; the upper bench showing a tendency to run into a block coal. The coal shows iron pyrites; and the parting slate is filled with it.

The bed above is only a two foot bed where opened in the same hill side.

Coal has been opened on Mr. A. H. Shaw's land, two-thirds of a mile east of Clearfield. The mine is now fallen shut, but is reported as having yielded as follows:

Shales, rusty,	25′	
Coal,	1	4″
Slate,	0	10
Coal,	1	8

"The coal hard, and burning freely in the grate." Some 80 feet above the last, a small 18 inch coal bed was opened, hard and bright coal, but too small to work.

The section (Fig. 42) shows the measures as exposed on a small run one-half mile east of Clearfield, as follows:—

Fig. 42.

Hill top.		
Shales,	30′	
Slate,	4	
Coal,	2	10″
Fire-clay,	2	
Shales,	25	
Coal, small.		
Concealed measures,	15	
Coal,	1	6
Concealed measures,	41	
Shales,	25	
Coal,	1	4
Slate,	0	10
Coal,	1	8
Concealed measures,	70	
Level of Susquehanna River.		

A. M. Hill's mine, two-thirds of a mile east of Clearfield, shows:—

Roof, black slate,	4	
Coal, block,	1	2″
Slate,	0	1
Coal,	1	8
Soft whitish fire-clay floor.		

The coal shows much pyrites and the slate is filled with it.

An average specimen of the coal gives, by analysis, (M'-Creath):

"Water,	0.380
Volatile matter,	22.280
Fixed carbon,	67.995
Sulphur,	2.455
Ash,	6.890
	100.000

Coke, per cent., 77.34. Color of Ash, dirty gray, with reddish tinge. The percentage of sulphur is heavy. The coal is shining, columnar, very friable, with much pyrites and charcoal in veins."

The section (Fig. 43) gives the measures exposed on the east bank of the Susquehanna river at Clearfield, just east of the south end of the town, as follows:—

Fig. 43.

Clearfield.

Hill top.	
Shales,	25′
Coal.	
Sandstone,	45′
Shales,	25
Sandstone,	7
Slate,	5
Coal,	2 6″
Fire-clay,	3
Shales, mainly,	83
Slate,	4
Concealed measures,	39
Bottom of hill.	

The only coal of this section now opened is Moore's coal, which shows roof grayish blue slate; coal 30 inches; fire-clay floor. The coal looks hard, bright, clean and good, and is well spoken of for house use.

West and north-west of Clearfield, the coals have been opened in many places for local use.

At Humphrey's Bank, one mile west of Clearfield, on Widow's Run, 195 feet above the Susquehanna by barometer, the coal shows:

Black slate roof.	
Bony coal and slate,	0′ 1½″
Coal,	2
Fire-clay floor,	1 6

The coal has been used in the Clearfield gas works.

A bench shows on the hill-side 45 feet above the coal, and a small bed was once opened in the valley below at water level, 50 feet below the mine.

A fair average specimen of Humphrey's coal gave on analysis, (M'Creath):

"Water, - - - - - - -	0.410
Volatile matter, - - - - - -	21.800
Fixed carbon, - - - - - -	72.903
Sulphur, - - - - - - -	1.087
Ash, - - - - - - - -	3.800
	100.000

Coke per cent., 77.79. Color of Ash, reddish.

The coal is bright, friable, fracture showing chisel faced forms. Pyrites in veins."

Mason's coal bank is opened about one-half mile west of Humphrey's, and on the upper bed. It measured:—

Black slate roof, - - - - -	4′
Bony coal, - - - - - -	1
Coal, - - - - - - -	2′ 4″ to 2 6″
Fire-clay floor, - - - - -	1 6

The coal shows some irregular small slate partings.

An average specimen of the coal from the upper part of the main bench, yielded (M'Creath):

"Water, - - - - - - -	0.550
Volatile matter, - - - - - -	22.650
Fixed carbon, - - - - - -	72.616
Sulphur, - - - - - - -	1.334
Ash, - - - - - - - -	2.850
	100.000

Coke per cent., 76.80. Color of Ash, red.

The coal is bright, columnar, containing veins of charcoal and pyrites."

An average specimen of the lower part of the main bench, yielded on analysis (M'Creath):

"Water, - - - - - - -	0.480
Volatile matter, - - - - - -	22.320
Fixed carbon, - - - - - -	59.788
Sulphur, - - - - - - -	4.232
Ash, - - - - - - - -	13.180
	100.000

Coke per cent., 77.20. Color of Ash, purplish.

The coal has a glossy lustre, is very friable and contains a very large amount of Iron Pyrites."

Such an analysis condemns entirely the lower part of the bed.

Another opening on the "Humphrey's Bed" made 600 yards east south-east of the first, showed about the same thing—two feet of good coal.

A section of the hill at this point gives:

Hill top.	
Shales, - - - - - - - - -	20′
Black slate outcrop.	
Grey and brown shale, and thin grey sandstones, the whole indistinctly seen,	100
Coal, - - - - - - - - - -	2 8″
Shales, - - - - - - - - - -	40
Coal, Humphreys, - - - -	2′ to 2 6
Concealed measures, - - - - -	30
Creek level.	

About one and a half miles north of Clearfield town, on the waters of Stone creek, Mr. J. Shaw has opened two coal beds, 55 feet apart.

The lower bed, 175 feet above the Susquehanna by barometer, measured:—

Black slate roof, - - - -	3′ or more.
Bony coal, - - - - - -	0 1″
Coal, - - - - - - - - -	2 3
Fire-clay floor, hard.	

An average specimen yielded on analysis, (M'Creath):—

"Water, - - - - - - -	0.520
Volatile matter, - - - - -	21.030
Fixed carbon, - - - - - -	67.133
Sulphur, - - - - - - -	.767
Ash, - - - - - - - -	10.550
	100.000

Coke, per cent., 78.45. Color of Ash, reddish gray.

The coal is bright, shining, very hard, with slate and charcoal in veins.

The upper bed, 230 feet above the Susquehanna River, by barometer, measured:—

Black slate roof, - - - - -	4′
Coal, hard, some bone mixed, -	0 6″
Coal, - - - - - - -	2 5
Fire-clay floor, hard.	

Some few non-persistent knife edges of slate run through the coal, which on the whole looks well.

About one-half mile west of these openings are the "Old Collins mines," also opened on a branch of Stone Run. Here the same two beds show, 55 feet apart. Between these beds the rocks are massive sandstone, fine grained, brownish, micaceous.

The lower bed measured: —

Black slate roof.	
Bony coal, - - - - - - -	0′ 3½″
Coal, - - - - - - - -	2
Fire-clay floor, hard.	

The upper bed measured:—

Black slate roof.	
Coal, hard, some slate mixed, - -	0′ 4″
Coal, - - - - - - - -	2 8
Fire-clay floor.	

The coal looks very well and does excellently for house use.

Five hundred yards south-west of these mines, R. Shaw has opened up this upper bed. It shows:—

Black slate roof,	
Bone coal, - - - - - - -	0′ 9″
Slate, persistent, - - - - - -	0 3
Coal, - - - - - - - -	1 7
Fire-clay floor, hard.	

An average specimen of this coal, yields on analysis (M'-Creath):

Water, - - - - - - -	0.870
Volatile matter, - - - - -	21.680
Fixed carbon, - - - - -	68.928
Sulphur, - - - - - - -	1.302
Ash, - - - - - - - -	7.220
	100.000

Coke per cent., 77.45. Color of Ash, pinkish.

The coal has a dull lustre, is columnar, very friable, iridescent, with Pyrites and charcoal.

The coal in this mine is cut down to a smaller size than in any other opening on the same bed examined. The coal beds of this section are singularly free from irregularities of floor or roof; carrying their regular, very moderate thickness of coal, and persistent slate partings for considerable distances almost unchanged. But in this mine the floor is very uneasy; rolling up at times two or more feet, and disturbing the coal everywhere in the mine.

The general dip seems to be locally to the south-west.

All these openings as described are on only two beds; the difference in their height above the river at the different openings being due to the slight south-east sinking of the measures at this point.

These beds, as well as those to be described opened higher up the Susquehanna river and over on Clearfield creek, though fairly good in quality, are all small. The country is cleared and thoroughly cultivated for miles in all directions, and the numerous openings made for local use and the trial openings are sufficient to render it extremely doubtful whether any one of the coal beds contained in the 400 feet of lower productive coal measures showing in this place can ever be expected to reach a marketable size over any considerable area. It would certainly require at least a clear 3 foot 8 inches to a 4 foot coal bed to compete with the basins to the eastward, and no such regular and even 4 foot bed has ever been opened in the vicinity of Clearfield, although some special mines may reach close up to it. Of course, there is a steady and increasing demand for coal for home use, and for this purpose these mines will continue to be worked.

The fire-clay which is opened at Clearfield, will be discussed in a separate chapter, in connection with the other fire-clay deposits examined in the First and Second Basins, the Sandy Ridge, the Blue Ball and the Woodland fire-clays.

The measures underlying the Lower Productive Coal Measures at Clearfield are given by the record of a boring for oil, made at that place a number of years ago. The boring records, as well as samples of the rocks passed through, are carefully preserved by Josiah W. Smith, Esq., of Clearfield, by whom they were kindly loaned to the Survey.

One small coal, six inches thick, is reported as having been passed through a short distance below water level. The record is:

Surface—
at 45′ deep, ferruginous sandstone.
62′, brown sandstone.
78′, light colored sandstone.
90′, coarse iron-stained sandstone.

96′, black slate mixed with sand.
100′, light colored, crumbly, iron-stained sandstone.
137′, soft grey slate.
150′, crumbly, iron-stained sandstone.
170′, white sandstone.
180′, greyish white sandstone.
200′ to 215′, dark colored slate.
233′, light grey sandstone.
288′ to 300′, clay slate.
320′, slate.
345′, red slate.
400′ to 412′, reddish slate.
450′ to 490′; light colored, greyish, slate.
Not recorded until at
855′ to 860′, light colored sandstone.

At between 400′ and 500′ below the surface weak brine was struck, and at between 700′ and 800′ strong brine was found. An analysis of this brine, by Prof. Geo. H. Cook, of New Jersey, gave: the copy furnished by Mr. Smith:—

"Common salt,	69.010
Chloride of Bittern, calcium,	25.090
Chloride of Magnesian Bittern,	5.900
	100.000
Water,	88.952
Salt,	7.591
Chloride Calcium,	2.767
Chloride Magnesium,	.655
Oxide of Iron,	.007
Silica and Earth,	.028
	100.000

A gallon of this brine weighs 8¾ pounds, and contains two-thirds of a pound of salt. 84 gallons contain 56 pounds of salt or one bushel."

Passing up the Susquehanna River and near the bridge over the River, 2 miles below Curwensville, about the largest and best coal bed in this division of the basin was opened many years ago by Mr. Reed. "It lies between 80 and 90 feet above the River, dipping at a considerable angle to the north-west. The coal bed is here 3 feet 6 inches thick. It forms an excellent fuel as it has but little sulphur. Like most of the coals of the First and Second Coal basins it affects a species of columnar structure, being traversed by innumerable vertical fissures which render it somewhat friable.

Prof. W. R. Johnson's analysis gives:—

Volatile matter,	27.000	in 100 parts.
Fixed carbon,	73.000	" "
Earthy matter,	5.300	" "

Coal columnar, cubical, brittle, jet-black, with great lustre."

Near Curwensville and below the town a coal bed 24 to 30 inches thick has been opened about 70 feet above the River. The next workable bed, the equivalent of Reed's Bed before mentioned, is perhaps 100 feet above the former. It has been opened both above and below Curwensville, on the higher river hills. Where opened about 1 mile south-east of the town the bed is 3 feet 3 inches thick, including a band of slate, 2 inches thick, and 9 inches from the bottom. Higher up the river it has been opened by Dr. Hoyt and others, from whose mines large quantities were sent down the river in former years. Underneath this coal is a bed of limestone 3 or 4 feet in thickness.

The following (Fig. 44) is a section of the coal strata in the vicinity of Curwensville compiled by estimation (from Rogers' Final Report):—

Fig. 44.

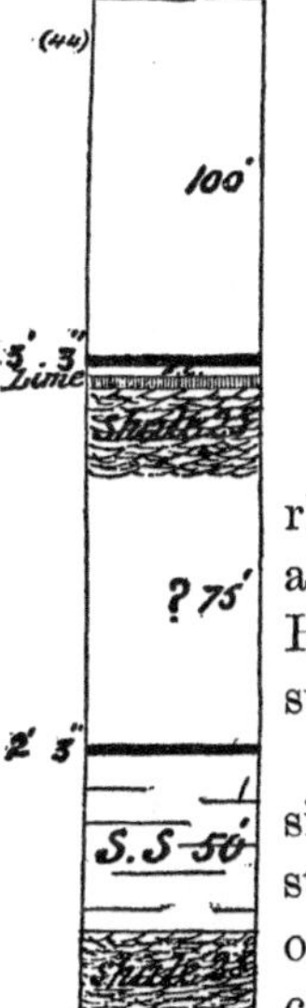

Hill top.		
Concealed measures,	100'	
Coal,	3	3"
Fire-clay,	2	6
Limestone,	3' to 4	
Shale, with two thin coal layers,	25	
Concealed measures,	75	
Lower coal,	2' to 2	6
Sandstone,	50	
Shales,	25 to river.	

One and a half miles above Curwensville, on the river, an old salt boring is said to have penetrated a six foot bed of coal, at a depth of 40 feet. Dr. Hoyt, who put down this well, makes the above statement.

One-half mile below Curwensville, on the north side of the Susquehanna river, 30 feet above the stream, there is an outcrop opening on a coal bed on the lands of Patton and Hoover. The outcrop shows:—

Black slate roof,	4'	
Coal smut,	0	6" to 8"
Fire-clay,	2	2
Coal smut,	1	1
Fire-clay floor, sandy,	2	6

Massive sandstone underlies the fire-clay for 10 feet. Thin bedded sandstones and coarse brownish and grayish shales overlie the coal for 80 feet to a small bench, then 90 feet of shales, with small pieces of sandy hematite through the mass, to an old opening, (now fallen shut,) on the "4 foot" bed of coal, then shales 20 feet to hill top.

These Curwensville coals are of excellent quality, but no bed rises above 3 feet 3 inches as far as opened.

On Anderson's Hill, about one and a half to two miles west of Curwensville, an impure ferruginous limestone shows along the main road. A fire-clay ten feet thick overlies it, and then a coal smut, called a "30 inch" coal.

Four miles above Curwensville, on the Susquehanna river, at J. Farwell's, is an exposure strongly resembling the above. The same impure limestone shows, with coal reported above and a yellowish and grayish sandstone underlying. As this exposure is near the centre of the synclinal axis, while Anderson's Hill is rising rapidly to the second anticlinal axis, the difference in level is easily explained.

Fig. 45.

Analyses of this so-called "carbonate ore" were made, (M'Creath,) and the specimens forwarded yielded respectively 4.80 per cent. and 8.00 per cent. metallic iron.

Going east from Clearfield on to Clearfield creek, Mr. Owens has made some openings to show the measures there. The section (Fig. 45) shows:—

Hill top	
Slate,	30′
Bench, coal?	
Slates, shales and thin sandstones,	30
Bench, coal?	
Slates and shales,	66
Limestone,	4
Concealed measures,	42
Black slates,	11
Coal,	2
Grey slates and shales	56
Coal, small,	2
Fire-clay, impure in parts,	15
Concealed measures,	105
Massive whitish and brownish granular sandstone,	40
Bed of Clearfield creek, sandstone.	

The mass of fire-clay, 15 feet thick, and in places 20 feet thick, is opened up and shows:

Coal, small, on top.	
Impure clay, - - -	3′ to 4′
Hard clay, - - -	5
Soft clay, - - - -	3
Hard clay, - - -	4
Soft clay, - - - -	6 to 8
Greyish clay on bottom.	

Where this fire-clay was opened up, some 600 yards south of Owens' house, a bed of coal, 50 feet above it, had been opened many years ago, but is now fallen shut. It is reported as a "5 foot bed." It is the bed which underlies the limestone in Fig. 45, and which is only two feet thick where opened near the house. The "5 foot" statement is probably exaggerated.

A specimen of this limestone from Owens' place was analyzed by Mr. M'Creath, at the laboratory of the survey in Harrisburg, and yielded:

"Carbonate of lime, - - - -	91.880
Carbonate of magnesia, - - -	1.892
Sulphur, - - - - - - -	.135
Phosphorus, - - - - - -	.031
Insoluble residue, - - - -	2.770

The limestone is dark blue and crystalline."

Rogers' Final Report mentions that on the north side of the Susquehanna river, three-quarters of a mile above the town of Clearfield, occurs a stratum of fire-clay of good quality, and eight feet thick. "Its position is a few feet above the river bank. Oolitic iron ore, seemingly of good quality, abounds in the upper part of the fire-clay, two feet of which contain about one foot of ore."

About 400 yards north of Clearfield station, on the Tyrone and Clearfield railroad, a well, going down at the time of the examination, gave the following:

Surface, - - - - - - - -	2′ to 3′
Small coal smut, - - - - -	1
Loose rocks and clay, - - - -	3
Black slates, - - - - - - -	3
Coal, - - - - - - - -	0 6″
Fire-clay, with numerous iron ore balls in the lower, 3′, - - -	8
Coal and slate, - - - - - -	1 6
Fire-clay, - - - - - - -	8 in shaft.
Coal, - - - - - -	6″ to 12 in shaft.

The beds are wavy and irregular, except the ore bearing fire-clay, which ran regularly as far as exposed.

A layer of similar Oolitic ore, supposed to be the same, exists on Clearfield creek near the water level, at the point of the Long Bend, amounting to nearly two feet of ore balls in close contact. In the steep hill above this bed several seams of coal have been opened by Joseph Irwin, and the following strata are developed, as shown in the accompanying vertical section, (Fig. 46):

Fig. 46.
Irwins

Hill top.		
Concealed measures,	20′	
Coal smut.		
Limestone, nodular.		
Slate, sandstone and shale,	100 or more.	
Coal,	1	6″
Fire-clay,	2	
Sandy shale,	25	
Compact, obliquely bedded sandstone,	25	
Slate,	25	
Coal,	0	8
Slate,	5	
Limestone,	6 to 7′	
Calcareous iron ore,	2	
Sandy shale,	60	
Coal, in two benches, the lower one pyritous,	4	6″
Fire-clay,	4	
Slaty sandstone,	100 or more.	
Coal,	1	6″
Slate, with iron ore nodules,	12	
Coal,	0	6
Slate,	3	
River at low water. (Rogers.)		

The only coal bed then opened up (it is not now worked) was the large bed beneath the thick limestone. The limestone is quite pure and burns to a good lime.

It may be mentioned that near this place on Clearfield Creek, Mr. Geo. Butler handed for examination a specimen of Cerussite, or carbonate of lead, with a trace of silver, which he reported as found in the neighborhood: but would not point out the place. It may also be mentioned that at two different places in the coal basins examined, specimens of imported Spiegeleisen were brought for examination with assurances of a regular bed existing close by.

A few miles farther up Clearfield creek, about one mile above where Little Clearfield creek comes in, on Lambert's place, is an exposure of a very peculiar ore deposit. The lower massive sandstones of the Productive Coal Measures, which have before been in the bed of the creek are here carried up well into the hills in obedience to the first anticlinal axis before mentioned, which passing somewhere near Stoneville, continues southwestward through Knox, Jordan and Chest townships. This axis has been already described as the dying to the north of the Great Laurel Hill axis. At Lambert's, therefore, massive sandstones, but without conglomerate layers so far as seen, make the sides of the creek for 100 to 125 feet above the water. Two hundred feet above the creek level, the surface of the ground is covered for some acres with a peculiar, rough looking iron ore, in lumps of all sizes, some of the pieces making from 150 to 200 pounds. In the midst of this outcrop, which fades away both on the ends and sides, but continues longest and most decided down the face of the hill, a shaft has been put down, which shows thus:—

Outcrop lumps on surface.	
Loose sandstone pieces, with some few ore lumps,	5′
Ferruginous sandstone with lean ore, - - - -	5
Red, clayey ore, - - - - - - - -	1′ to 2
Clay slate, ferruginous, with some red ore, - -	6
White sandstone in bottom.	

A specimen of the best quality of the outcrop surface ore yielded on analysis (M'Creath):—

Iron, - - - - - - - -	42.400
Sulphur, - - - - - -	.039
Phosphorus, - - - - -	.082
Insoluble residue, - - - -	23.120

The above analysis represents an iron ore of very good quality, but the great mass of the ore deposit was leaner and more sandy.

The ore analyzed was a limonite, compact, with laminated structure, and reddish brown color.

Thirty feet above this iron ore there is the smut of a small coal, the lowest coal bed showing in these measures.

About 400 yards up a small run which enters Clearfield creek at this point, a bog ore is found, extending apparently over some acres. As the stream bed here is at the bottom of the seral conglomerate, the bog ore may represent iron brought from the carbonate iron ore of XI, just below the conglomerate of XII. The ore deposit seemed to extend over an area of about one hundred yards in length along the valley, by about fifty yards in breadth; the depth was not found, but apparently it was not great.

CHAPTER XVIII.

Description of the Coal District of Clearfield Creek, in the First Bituminous Coal Basin of Clearfield County, Pa.

Clearfield creek, after cutting through this First or Laurel Hill anticlinal axis at Lambert's, comes again into productive coal measures, and runs through them up to, and slightly beyond Madera.

An examination of the vertical sections already given in describing the country from Karthaus to Clearfield creek, shows many points of difference. Without entering into these different points, which more properly comes under a discussion of thinning and thickening of the measures and beds of all the basins, it is sufficient to indicate that the limestone near the hill top at Karthaus and on Clearfield creek, may be sufficently identified with the Freeport limestone: the Upper, Middle and Lower Freeport coals appear, although in a sadly attenuated condition, and the coals below the Freeport group are shown in the openings around Clearfield.

In going westward from Osceola to Madera, across the first anticlinal sub-axis, the heavy outcrop of massive conglomerate, with white rounded quartz pebbles, ranging in size from a pea to an olive or an egg, marks the centre ot the anticlinal sub-axis at about Haggerty's, a little east of Amesville, in Woodward township. Before reaching this axis, while still in the eastern sub-division of the First Basin, coal bed B is found opened at M. Walker's four miles west of Osceola, reported "3 feet 10 inches thick," the mine now fallen shut. Two and a half feet of black slates overlie the bed, and on top of them six feet of rusty, dark colored slates. Another bed was opened below this one, in the ravine, many years ago. It is reported a large bed, and would indicate that coal bed A also possesses value through this region. At A. Walker's place also, one-half mile beyond, the so called "big bed" or "six foot" bed is reported as once opened up in full thickness, with two small beds above, making it probably bed B also in this case. After crossing the anticlinal sub-axis and going west from Amesville to Madera, small black slates and coal smuts appear on the road side, markedly at C. C. Shoff's, and near the school house; but no development is found until reaching Madera.

But the anticlinal sub-axis which has come south-westward from the Blue Ball in a very straight line is here materially changed, or rather dies down gradually and is re-placed by another axis just east of it. Running south-westward from Amesville, it is seen in Becaria township, west of Utahville, as a very flat dying axis, carrying on its back the soft shales and slates above the conglomerate. But at Janesville, due south of Amesville five miles, and therefore east of this dying anticlinal axis, the massive seral conglomerate shows very plainly on the crest of a well marked anticlinal. These axes are therefore *en echelon;* one extending north-eastward as the vigorous axis already described, and the other going south-westward into Cambria county.

Between Janesville and Amesville, the lower beds of the Lower Productive Coal Measures come in, and coal has been opened at M'Cully's and Whitesides' places. The opening on the road-side, (M'Cully's,) was filled with water, and only two feet of coal could be seen.

Two miles south-east of Janesville, and of course in the Eastern sub-division of the First Basin, the continuation of the Osceola and Beaver Branch Coals, coal has been opened and is worked on a small scale at the G. W. Davis mine. The coal as measured in the mine shows:—

Fig. 47.

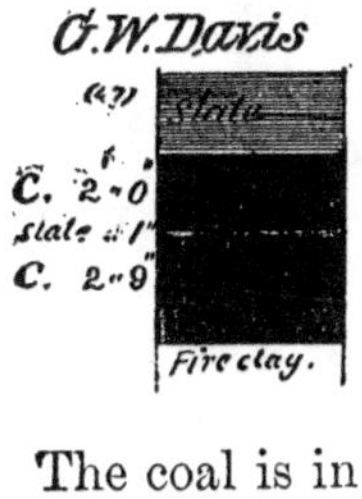

Black slate roof.

Coal, - - - - -	2′	
Slate, - - - -		1″ to 2″
Coal, - - - - -	2 9	

Fire-clay floor.

The coal is in all probability Bed B of the Osceola section.

A specimen of the coal forwarded for analysis yielded as follows, (M'Creath):—

"Water,	0.640
Volatile matter,	23.010
Fixed carbon,	71.799
Sulphur,	.551
Ash,	4.000
	100.000

Coke, per cent., 76.35.

The coal has a dead lustre, is friable, contains charcoal and Pyrites in veins, slightly iridescent."

The above analysis represents an excellent coal: giving this bed through this region both size and good character.

Fig. 48. *Madera*

At Madera, on Clearfield Creek, and in the Western sub division of the First Basin, a section of the hill side (Fig. 48) gives the measures exposed as follows:—

Hill top.	
Shales,	20′
Small bench and coal smut.	
Shales,	55
Bench, reported once opened a 6′ coal.	
Thin sandstones and slates,	40
Bench, not opened.	
Shales, buff and brown, with a little lean hematite iron ore in small pieces,	31
Fire-clay, not opened, reported,	9
Black slate,	5
Coal, not opened, called	5
Shales and slates,	41
Slates, hard and dark colored, with nodular carbonate iron ore,	9
Black slate,	5
Coal, not open, called	4
Sandstone,	26
Coal, called	4
Sandstone,	30
Level of Clearfield Creek, coal reported in creek bed.	

Not one of the coal beds on the above section is now opened and worked. Mr. Jobling reports that he opened up the beds at 30 and 60 feet above the creek: the lower one four feet and the upper one six feet two inches thick. But all these openings are now shut, and no such large and valuable beds were seen. In all the openings around Madera, a gentle dip to the south has been found.

Luther's opening on the bed, 30 feet above the creek, is also fallen shut and cannot be measured.

Mr. D. W. Blair, of Pottsville, states that in a boring made at Madera the show was:—

Surface—
At 33′, micaceous sandstone.
48′, hard white flinty sandstone.
110′, pieces of bituminous coal.

The rock coming up with this coal appeared to be a brownish conglomerate sandstone.

North-west of Madera, on the opposite side of Clearfield creek, Mr. Haggerty has opened up a coal (Bed B probably) 80 feet above the water. The coal measured in the opening:

Fig. 49.
Haggertys' Bk.
C....2′ 4″
Slate.... 3″
C... 2′ (?)
Fire clay.

Black slate roof.

Coal, - - - - -	2′	4″
Slate, - - - - -		3 to 4″
Coal, partially covered up,	2	

Fire-clay floor.

Some bastard limestone, very sandy, lying outside of the mine, but not seen in place, seemed to have come from below the coal. A massive sandstone, slightly conglomeritic, underlies this coal. The dip, locally, is very slight to the south-east, (almost horizontal,) and the coal lies, therefore, slightly higher in the hill than where opened up at Madera. Going west on the road from Madera to Glenn Hope, a coal smut shows on the road side near Shoff's house, half a mile west of Madera, about 80 feet above the creek. Between Madera and Glenn Hope limestone has been quarried on the bank of Clearfield creek, about 40 or 50 feet above water level; a small coal smut showing also 20 feet above the water.

Fig. 50.
Glenn Hope

Coal bench
Coal bench
Coal bench

The measures show but imperfectly at Glenn Hope; as seen in the hill, back of the village, the section gives (Fig. 50) as follows:

Hill top.

Shales, - - - - - - -	14′	
Black slate, - - - - - -	4	
Coal, - - - - - - -	2	8″

Fire-clay, - - - - - -	2′
Grayish slates - - - -	12
Limestone, - - - - - -	6
Shales and slates, - - -	49
Bench, coal?	
Thin slaty sandstones and some gray slates, - - - - - -	105
Bench, coal?	
Concealed measures, - - -	35
Bench, coal?	
Slaty sandstones and shales, -	95
Level of Clearfield creek.	

Only one coal is opened in this hill side, and that is the one above the limestone, and is worked by Dr. Caldwell: it yields but 32 inches of coal. The coal is hard, good looking, and breaks out in blocks. The limestone is mainly blue in color, with some grayish layers, and burns well for lime for agricultural purposes.

A specimen of this limestone from Caldwell's quarry, forwarded for analysis, yielded, (M'Creath):

"Carbonate of lime, - - - -	93.810
Carbonate of magnesia, - - -	1.710
Sulphur, - - - - - -	.053
Phosphorus, - - - - -	.008
Insoluble residue, - - - -	2.070

The limestone is hard, compact, crystalline, and of a bluish gray color."

This limestone can be identified as in all probability the Freeport Limestone, and the small overlying coal as the Middle Freeport Coal Bed.

On the east side of the creek, Mr. J. Cooper has opened up one bed and is working it. The coal is small, not averaging over 30 inches in thickness. It is fairly good looking coal and answers satisfactorily for local use.

A specimen of the coal forwarded for analysis, yielded, (M'-Creath):

"Water, - - - - - -	0.700
Volatile matter, - - - - -	24.020
Fixed carbon, - - - - -	64.951

Sulphur, - - - - - - -	1.639
Ash, - - - - - - - -	8.690
	100.000

Coke, per cent., 75.28. Color of Ash, red.

The coal has a dull lustre, showing much oxide of iron, very hard, with veins of pyrites and charcoal."

A vertical section made near this mine, about one mile south-east of the village, shows (Fig. 51) as follows:

Fig. 51.

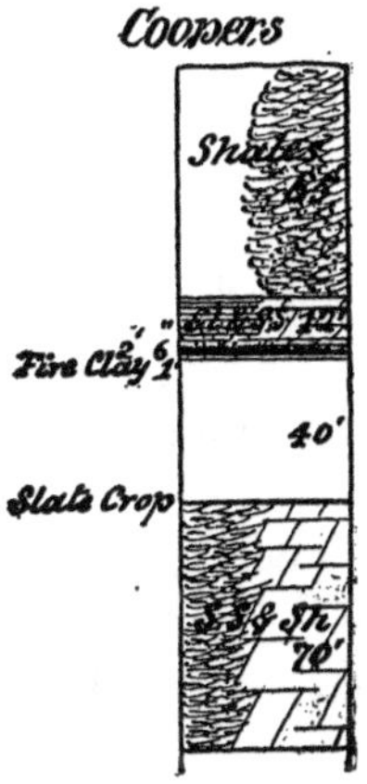

Hill top.

Shales, and some slates, - - - -	65'	
Grayish slates and thin sandstones, -	12	
Sandstone, - - - - - - - -	1	
Coal, - - - - - - - - -	2	6"
Fire-clay, - - - - - - - -	1	
Concealed measures, - - - - -	40	
Small black slate outcrop.		
Slaty sandstones and shales, - -	70	
Level o: Clearfield creek.		

At the south-east end of Glenn Hope village a fire-clay shows on the road side, not far above the creek. The bed of Clearfield creek for some miles above Glenn Hope, is made up of massive sandstones, usually grayish. At the bend in the road, near the school house, one and a half miles south-west of Glenn Hope, the following section shows on the west side of the creek (Fig. 52):

Fig. 52.

Hill top.

Shales, - - - - - - - - -	20'
Bench, small.	
Soft shales, - - - - - - - - -	45
Bench, coal?	
Thin bedded sandstones and grayish slates,	55
Black slate.	
Shales and slates, - - - - - - -	35
Conglomeritic sandstone, - - - - -	40
Coal smut.	
Sandstone, - - - - - - - - -	35
Level of Clearfield creek.	

At the mouth of Witmer run fire-clay shows in the bed of Clearfield creek.

At J. Leightner's place, on the east bank of Clearfield creek, one mile above the mouth of Wit-

Fig. 53.
Leightners'

mer run, the following measures are exposed (Fig. 53)

Hill top.	
Shales,	15′
Lean hematite ore in shales.	
Bench, small.	
Shales,	70
Coal, called	"3"
Limestone crop, sandy and poor.	
Gray slates,	38
Coal.	
Thin sandstones and shales,	46
Coal and fire-clay partings, called,	9
Brown shales,	12
Iron ore, carbonate, in layers,	"3"
Shales,	15
Sandstone,	15
Slate,	5
Coal, called,	5
Fire-clay,	8
Sandstone, massive,	12
Level of Clearfield creek.	

The fire-clay exposed near the bottom of the section is rather sandy. The coal, 20 feet above the creek, is only partially opened up on the outcrop, not sufficiently to verify Mr. Lightner's statement of a "5 foot bed." The iron ore, 60 feet above the creek, is not opened out fully, but shows three layers of carbonate iron ore of six, three and four inches respectively, making 13 inches in all. It is reported to have shown as a "3 foot" bed of ore when fully exposed.

A specimen of this carbonate iron ore of J. Leightner's, forwarded for analysis, yielded (M'Creath):—

Iron,	34.000
Sulphur,	.061
Phosphorus,	.356
Insoluable residue,	18.050

The ore is a carbonate, minutely crystalline and of a dark grey color, with conchoidal fracture.

The above analysis shows a very fair iron ore.

The outcrop of coal, slate and fire-clay, 75 feet above the creek, was not in condition to permit of accurate measurement. Mr. Leightner's statement is that the bed showed when opened up on the crop by a trial pit:

Black slate roof	
Coal, - - - - - - -	5′
Fire-clay parting, - - -	1
Coal, - - - - - - -	3
Fire-clay floor.	

The outcrop and bench are decided: but not so large as to indicate the presence of any such great bed as named above. As the certain presence of a regular five foot bed would be of great importance in expediting the opening of this region to market, this bed should be proved in many places. It lies here low in the hill and covers a considerable area.

The outcrops and benches above this bed have not been opened except the "3 foot" coal bed, over the limestone, which was opened and worked, but is now fallen shut. It is said to have yielded a hard, bright, clean coal, strong burning and used raw by blacksmiths. The limestone is placed in the section from some few outcrop pieces of bastard limestone on the surface. A limestone, however, is worked, about one mile south-west of Leightner's place, and is four to five feet thick: and also on the west side of Clearfield creek, high in the hill tops, about at the level of this outcrop.

The hematite ore in the shales at 240 feet above the creek level, is small in quantity and usually decidedly lean and poor in quality, though occasionally masses of very fair ore are found: and about the same thing may be said of every other crop of hematite ore in shales observed in these Lower productive coal measures during the season's work.

A specimen of the better quality of this hematite iron ore from the shales, forwarded to Mr. M'Creath for analysis, yielded:—

Iron, - - - - - - - -	40.800
Sulphur, - - - - - -	A trace.
Phosphorus, - - - - -	.596
Insoluble residue, - - - -	25.600

The ore is a limonite, hard, compact, silicious, and of a dark brown color.

South of Leightner's, the beds of coal shown in the section (Fig. 53) have been opened and worked in various places in the past: mainly small openings, worked only for a part of the winter to supply the trifling local demand, and abandoned.

No measurements could be made. But from the statements made it seems in every way probable that one, if not two, good workable coal beds spread through this southern part of Beccaria township on both sides of the valley of Clearfield creek.

It must be remembered that these sections at Glenn Hope and at Leightner's on Clearfield creek are in the Western sub-division of the First Coal Basin: occupying geologically the same position as Karthaus to the north-east of them, and of Johnstown to the south-west of them.

The examination was not carried southward into Cambria county, nor westward on to the waters of Chest creek. The divide between the waters of Clearfield creek and Chest creek represents the centre of the dying First anticlinal axis before described.

Fig. 54. Turner's Run.

Coming eastward from where Clearfield creek passes the Cambria county line, in Beccaria township, towards Utahville, the following section is found on Turner's Run:

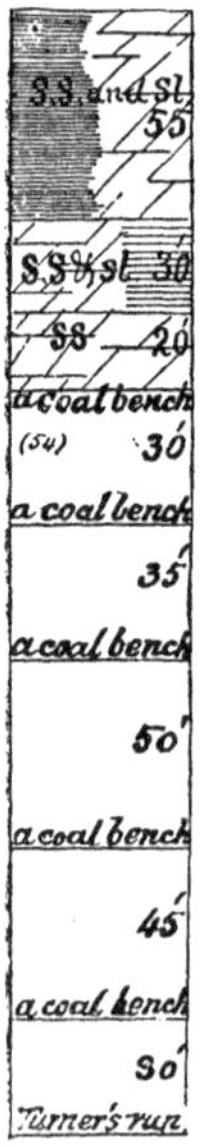

Hill top.	
Thin sandstones and slates, - -	55′
Small black slate show.	
Thin sandstone and grey slates, -	30
Sandstone, - - - - - -	15
Bench.	
Concealed measures, - - -	35
Bench.	
Concealed measures, - - -	35
Bench.	
Concealed measures, - - -	50
Bench.	
Concealed measures, - -	45
Bench.	
Concealed measures, - - -	30
Level of Clearfield creek.	

The highest point on the road between the waters of Clearfield creek and those of Muddy run is near Hopkins' house, 375 feet (by barometer) above Clearfield creek. This represents apparently the dying down of the First Anticlinal sub-axis which has been mentioned already as coming straight from the Blue Ball to Amesville, to die out between Muddy run and Clearfield creek. West of Hopkins, and 70 feet lower, a small black slate show is seen on the roadside.

East of the crest of the divide a small coal show is found at Flick's, 155 feet (by barometer) below the highest point. From Utahville to Muddy run the same small black slate crops are found, and the run is found cutting deep, being only 15 feet higher (by barometer) than Clearfield creek at the mouth of Turner's run.

Going eastward up the hill which divides the Muddy and Little Muddy runs, black slate crops show at 20, 50 and 75 feet above the water; thin bedded sandstone at 110 feet; and again black slate near the Keystone Inn, at 165 feet above Muddy run. The valley of the Little Muddy, (which does not cut so deep as Muddy run by 75 feet,) is filled with masses of conglomerate, with large white rounded quartz pebbles; and this same massive coarse conglomerate is found at Janesville, 50 feet above the stream.

In describing, in a previous chapter, the general geology of the county, the position of affairs at this point, with its anticlinal axes *en echelon*, was fully stated.

The proper relationship of these coals of the western sub-division of the First Basin to those of the eastern sub-division, and other basins, and of the coals of the different vertical sections to each other also, will be discussed in a review of all the sections of the past season. Some few points show even at a glance. There is much more limestone and carbonate iron ore in the western than in the eastern sub-division of the First coal basin.

At Osceola and vicinity, in the eastern sub-division of the First Basin, there are three coal beds of workable size, A, B and D; and two of them of excellent character, B and D.

At the Snow Shoe, also in the eastern sub-division, there are four beds of workable size, A, B, D and E; and two or even three of good character.

At Karthaus, in the western sub-division of the First Basin, there is an abundance of coal, the beds of workable size and good quality, and also much iron ore; and at Glenn Hope and Leightner's, along Clearfield creek, also in the western sub-division, the sections show the presence of workable coal beds and of iron ore.

Though the coals of these sub-basins above named come over

into the Second Bituminous Coal Basin across the broad and gently rolling arch of the first axis, yet the vertical sections of the Second Basin show coals smaller in size and inferior in quality; and also much less iron ore.

The fire-clay deposits are large and good, both in the First and Second Basins.

CHAPTER XIX.

Description of the north-west part of Clearfield county, in the Second and Third Basins.

Going north-west from Curwensville the limestone and lean carbonate iron ore crop is found at Holden's, high up on Anderson's hill, 400 feet above the Susquehanna River, with the small coal smut showing just above it, as already described. Another coal smut shows 65 feet above the limestone and a fire-clay at 85 feet above. Above this, thin sandstones and shales to the hill top, 55 feet, with lean brown hematite showing in small quantities in the shales, which are very rusty. The hill top is about 500 feet (by barometer) above the Susquehanna River. Black slate shows on the road side west of Derrick run, and fire-clay outcrops at J. Ellinger's: but from this point on, the exposures are sandstones, massive, fine grained and occasionally conglomeritic in small grains, but without showing any layers of white quartz pebble conglomerate. The centre of the Second anticlinal axis is passed near Bloom's Run: for west of that place the sandstones, although almost horizontal, give a slight sinking to the north-west. At Packersville the whole surface is covered with small sandstone lumps, whitish and fine grained. The soil and roads are very sandy.

Fire-clay shows in the road about half a mile west of Packersville, and a small black slate shows not far beyond. From this on to Luthersburg the outcrops of the black slates and clays are seen repeatedly on the road side as it crosses the hills and valleys.

The measures around Luthersburg have been opened in

numerous places: and the openings confirm the section in the Rogers' Final Report, which is as follows, (Fig. 55):

Fig. 55.

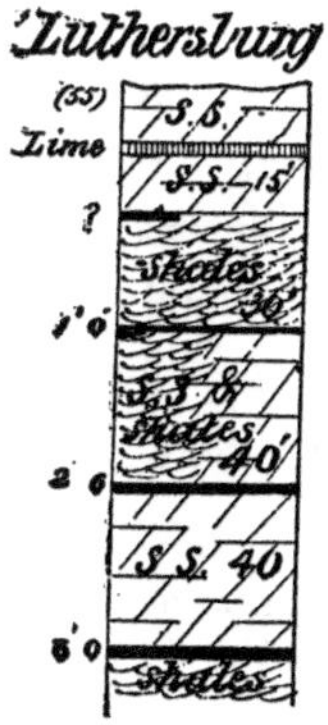

Luthersburg

Hill top.		
Sandstone,	14′	
Coal?		
Limestone,	3	
Sandstone,	15	
Coal.		
Shales,	30	
Coal,	1	
Shales and sandstones,	40	
Coal,	2	6″
Sandstone,	40	
Coal,	3	
Shales.		

In scarcely any case where these coals have been opened has the thickness exceeded three feet. This of course restricts the production to the amount needed to supply the local demand.

Two coal beds are developed north-east of the town. The upper one measures two feet six inches and the lower one two feet six inches to three feet. The lower coal is mined near the Clearfield turnpike, three-fourths of a mile east of the town, where it is two feet eight inches thick. Forty or fifty feet above it, on the turnpike, indications of the upper bed are seen, resting on fire-clay. The same bed was penetrated in a well on the hill-side at Luthersburg and proved to be two feet six inches thick. A coal bed one foot thick was cut through by several wells in the higher ground at Luthersburg. The existence of a fourth bed appears to be indicated by springs and black dirt on the hill just east of the town. It gives no evidence of being large. (Rogers' Final Report of Pennsylvania.)

About this locality, on the Curwensville turnpike, limestone has been quarried; it is also opened on the place of A. Pentz, Jr., one-half mile north-east of the town. It occupies only the high ground, and is three to four feet thick, gray, compact, sonorous, and when weathered, yellowish or brownish, from the amount of iron which it sometimes contains. It is burned somewhat for agricultural purposes, but will not make a white lime for plastering. These same measures spread north-east and south-west very evenly and regularly, a limestone being found in the hill tops, and burned for lime, four miles north-east of Luthersburg.

South-west of Luthersburg, the coal smut outcrops of these beds show repeatedly on the roadside between that place and Troutville.

The vertical section (Fig. 56) is compiled from observations made on the East Branch of Mahoning creek, one and a half to two miles east of Troutville. The section is necessarily compiled from scattered points, and is only approximately correct. It shows:

Fig. 56.
Troutville

Hill top.		
Concealed measures,	14	
Bench.		
Micaceous sandstone and *concealed measures,*	49	
Black slate,	3	or more.
Coal, Kopp's,	3	6″
Concealed measures,	73	
Bench.		
Concealed measures,	101	
Slate,	4	
Coal, Peacock, called,	3	
Gray slate,	9	
Fire-clay,	1	
Rotten gray slate,	10	
Coal, "under creek."	1	8

The coal under the stream was covered over at the time of the examination, and could not be measured. It is reported as a "20 inch" coal, and that a blue carbonate iron ore was opened under it.

The coal 20 feet above the creek level has been opened but is now fallen shut. It is a fair looking, Peacock coal, and is reported by Mr. Ginter as having shown a full "3 foot" coal bed, which burned well for domestic use. Massive sandstone lumps fill the valley of the East Mahoning creek, at this point.

The next bench above is not opened. Kopp's mine on the bench still farther above, is two miles east of Troutville. It measured (Fig. 57):—

Fig. 57.
Kopp's Bk.

Sandstone.	
Black slate roof.	
Coal,	3′ 6″ to 3′ 8″
Black slate in floor.	

The mine was partially filled with water and could only be imperfectly examined near the entrance. It is stated that the coal measured four feet six inches at the head of the main entry.

The present bottom as seen does not look like the true floor, and the additional thickness is every way probable. The roof is good, tough and dry. The coal was quite tender and friable where examined near the mine mouth; it is said to have been harder when under heavier cover. Some slate was found in irregular layers, but without any persistent parting. But little pyrites showed in the coal, and in every way it gave evidence of being a valuable coal bed.

The hill rises to the next bench, 55 feet higher up and not opened at all, and then 14 feet more to the hill top. As showing the local variations in dip which may be looked for in mining these flat coals, it may be stated that this Kopp mine is dipping to the south-west and south decidedly, while the regular dip of the measures at the point is about north-west, and this abnormal dip is continued as far as the mine has been driven, (about 80 yards.)

At E. Luther's natural opening one and a half miles east of Troutville, one-half mile above Pentz's mill, and thirty-five feet above the creek level, the coal measured:—

Grey slates and dark slates,	3′	
Black slate roof, thin bedded,	2	6″
Coal,	1	8
Floor, not seen.		

The coal seemed to dip to the south-west. The hill rises 80 feet above, but shows no marked bench.

On the Bell and Kramer property, one and a half miles south south-east of Troutville, coal has been opened up, 69 feet (by barometer) above the creek. It measured as reported (for it has now fallen shut):—

Black slate roof.	
Coal,	2′ to 2′ 1″
Slate,	1 4
Fire-clay, tough, plastic, thick.	

The coal is rising to the north-west.

Another bench is found at 136 feet above the creek, where, with a pile of black slate thrown out from a trial pit, there are numerous pieces of blue carbonate iron ore. From the appearance of the pieces, the carbonate ore must be bedded in thin plates in the slates. The hill side rises one hundred feet higher, but without any marked bench. South of Troutville, on the Godfrey Weaver place, in Brady township, a bed of limestone is exposed. It is a fair looking, dove colored limestone, and from its appearance when weathered, evidently contains considerable iron.

Going south-east from Troutville, the outcrops of the beds given above are found in several places, but in all cases appearing to be small. At Weaver's place, at the north end of Bell township, a small ten to twelve inch coal, lying upon a fire-clay floor, is opened just above water level. Some 50 feet above this, a ten inch coal was opened, and a small bench shows about 50 feet above this latter, but has never been opened up. Pieces of a bastard, sandy limestone are found in the stream wash. One-third of a mile south-west of Weaver's, in Bell township, Mongold's coal mine is opened and worked. It shows (Fig. 58):

Fig. 58.

Mongold's Bank

Roof clay slate, running into black slate,	6′
Bony coal and slate, - - - -	0 ½″
Coal, - - - - - - - -	1 11½
Black slate, persistent, - - - -	½
Coal, - - - - - - - -	1 7
Fire-clay floor, hard.	

The coal mines out hard and bright. An average specimen yields on analysis (M'Creath):

"Water, - - - - - -	0.860
Volatile matter, - - - - -	31.600
Fixed carbon, - - - - -	61.662
Sulphur, - - - - - -	2.228
Ash, - - - - - - -	3.590
	100.000

Coke, per cent., 67.54. Color of Ash, brown with red specks. The coal is bright, with shining lustre, rather compact, shows little pyrites."

The coal dips decidedly to the north-west.

This coal mine is high up on the Second Anticlinal axis, which probably has its centre not more than from one-half mile to one mile south-east of the opening. It represents, therefore, one of the lowest, if not the lowest workable coal bed of the Lower Productive coal measures, most probably coal bed A. Massive sandstone boulders, granular, fine grained, but not pebble conglomerate, lie all around the vicinity of the mine, and the seral conglomerate sand rocks are therefore the surface rocks south-east of the mine, to and across the Second Anticlinal axis.

It is plain, therefore, that this part of the Third Coal Basin, as at and around Luthersburg, yields only coal beds of moderate size and quality; and at present only available for local use.

Going north-west from Troutville to Stump creek, small black slate outcrops and benches mark the place of these small slaty coals of the Third Basin. North of where the Brookville pike crosses Stump creek, a decided bench shows near Reed's, 150 feet (by barometer) above the creek. The hill rises 100 feet higher, made up of gray slates and shales, but no other marked bench shows.

On a branch of Sugar Camp run, an affluent of Stump creek, three and a half miles south of the Sprague mine, Mr. Brown has recently opened a coal bed just above water level. The opening shows (Fig. 59):

Fig. 59.

Fire-clay roof.		
Rotten bone coal and slate,	1′	2″
Fire-clay,		4
Soft bone coal and slate,		8
Coal, hard,	3	8
Bone coal,		2
Fire-clay floor, hard.		

An average specimen of the coal yielded on analysis (M'-Creath):

"Water,	1.010
Volatile matter,	27.790
Fixed carbon,	48.365
Sulphur,	3.885
Ash,	18.950
	100.000

Coke per cent., 71.20. Color of Ash, gray with pink tint.

The coal has a dull lustre, is slightly iridescent, with a large amount of pyrites."

The opening, when examined, had only been driven in eight feet on the outcrop of the bed. A small black slate, one-half inch thick, shows about the middle of the main coal bench; and looks as if persistent. It is possible that on driving further in, on to more solid coal, some of the weathered bone coal of the section may prove to be workable coal; but the roof, though rather an abnormal one, (fire-clay) seemed to lie regularly in place. It will be noticed that the coal shows an enormous percentage of ash; partially due, no doubt, to the nearness to the outcrop. The mine lies just across the western line of Clearfield county, and is in Jefferson county.

The hill rises 128 feet above the coal, made up of grayish slates and shales. Though but little sandstone shows in the stream beds, yet this coal seems clearly to belong to the base of the Lower Productive Coal Measures.

A carbonate "iron ore," opened up slightly near Shaefer's house, proved too poor in iron to be worth working. It overlies the coal 30 feet. A small limestone bed is reported as once opened in the stream bed below.

The waters of Stump creek, which flows into Mahoning creek, cut deep, the level of Stump creek high up on its waters where crossed by the Brookville pike being 50 feet (by barometer) lower than the Sandy Lick creek at Reynoldsville. The difference in level between the Sandy Lick creek at Reynoldsville (1,330 feet of tide) and the Mahoning creek at Punxatawney, (1,142,) all in Jefferson county, is 188 feet.

The hills on both sides of Stump creek, as it goes south-west down the Third Basin, are low and rise back very slowly, taking in, therefore, but little of value.

The iron ores of these measures, whether occurring as hydrated sesquioxides or limonites, or as blue carbonate of iron or siderite, have been mentioned wherever they showed sufficient quantity to give any prospect of workable value in the future.

It has not been deemed necessary to mention specially all the cases where outcrops of nodular iron ore in shales or fire-clay were examined, nor the outcrops of lean, sandy hematite in shales.

These nodular deposits, though at times fairly continuous for considerable distances, do not afford sufficient hope of regularity of deposit to justify a regular working of them alone. In the case of the "Clarion ore," the clays over the ore have in them a very varying percentage of ore balls: and in driving the gangways through the overlying clay, keeping the 6 to 14 inches of solid ore in the bottom of the gangway, the mines at times find nodular masses almost sufficient to pay the entire expense of mining. But this is rare, and the operation is so uncertain that it is safe to say that there is scarcely to-day in Pennsylvania a single nodular iron ore deposit being worked alone, even on a small scale.

It should be mentioned that in the shales overlying the Dixon Mine, near Evergreen station, on the Bennett's Branch Division of the Allegheny Valley railroad, a fine red hematite shows in small pieces in shales. It analysed (M'Creath) 57 per cent. of metallic iron. But there was no evidence that it existed in any quantity.

It may be noted that Prof. T. Sterry Hunt, in his report on the Hocking Valley, (Ohio,) reports red hematite as showing in marked quantities in three places: As nodular masses on the surface, apparently above coal 8 of the Ohio system, overlying the Mahoning sandstone; in masses on the surface, between coals 6 and 7 of the Ohio system, probably just over the Mahoning sandstone; and thirdly a stratum of two feet, charged with nodules of it, about 30 feet above coal 1 of the Ohio system, or at the base of the Lower Productive coal measures of Pennsylvania. An analysis of one specimen yielded 61 per cent. of metallic iron and only 2 per cent. of volatile matter.

The coals opened in the extreme Western edge of Clearfield county, on the Bennett's Branch railroad, near Evergreen station, at Bell's, Rumbarger's Dixon's and other mines will be included in the report of the Reynoldsville Gas Coal Basin and a section of the Bennett's Branch road which geographically belong to it.

CHAPTER XX.

On the Clearfield County Fire-clays.

The fire-clays of Clearfield county are opened and worked in the First and Second Coal Basins, along the line of the Tyrone and Clearfield railroad, at Sandy Ridge, Blue Ball, Woodland and Clearfield.

The Sandy Ridge Fire-clay mines and works are at Sandy Ridge Station, on the Tyrone and Clearfield railroad, one mile north-west of the summit of the Allegheny mountain. Where measured, the clay showed (Fig. 60):

Fig. 60.

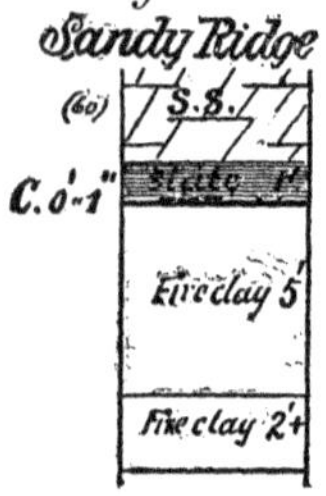

Sandstone.		
Black slate, - - - -	1'	
Coal, - - - - - -	0	1'
Fire-clay, worked, - -	5	
Fire-clay, hard, not worked,	2 or more.	

The hard fire-clay layer which is worked, ranges usually from 4 feet to 6 feet thick, averaging 5 feet or more; but ranges in places from 4 feet to 12 feet in thickness.

A small coal bed one foot thick is reported as showing 25 to 30 feet above the black slate layer of Fig. 60.

The clay worked is in three layers, and these are kept separate, the different qualities of these layers making them specially valuable for different purposes. The top layer is said to be adapted for furnace bottoms; the middle layer, the hard clay, is used for bricks, and the third layer for making tiles and the in-wall of furnaces. The hard sandy clay in the bottom is not worked.

Average specimens of these four layers of clay, yielded on analysis, (M'Creath):

1. Top layer. 2. Second layer. 3. Third layer. 4. Bottom layer.

	1.	2.	3.	4.
Silica, - - -	45.650	44.950	45.820	74.950
Alumina, - - -	34.730	37.750	35.950	15.940
Oxide of iron, - -	3.546	2.700	3.330	1.899
Lime, - - -	.112	.302	.112	.106
Magnesia, - -	.619	.216	.573	.407
Alkalies, - -	5.750	.985	4.130	1.756
Water, - - -	9.650	13.050	10.130	4.885
	100.057	99.953	100.045	99.943

A glance at the above table of analyses shows why the bottom layer is not worked.

An analysis of Sandy Ridge fire-clay made in 1870, by Mr. M'Creath, for the Pennsylvania Steel Company, and published by permission of the President of that company, yielded:-

Silica, - - - - - - -	45.880
Alumnia, - - - - - - -	33.920

Oxide of iron, - - - - -	4.680
Lime, - - - - - - - -	.160
Magnesia, - - - - - - -	.750
Alkalies, - - - -	4.643
Water, - - - - - - -	10.370
	100.403

The above analysis agrees very closely with the average of the three analyses of the top, second and third clay layers.

"Clay No. 1 (top layer) is massive, of pearl color, has a soapy feel, and the outside of the lumps slightly fibrous.

Clay No. 2 (second layer) is compact, massive, has grayish color, with bluish tint on fresh surface.

Clay No. 3 (third layer) is compact, of pearl gray color, and breaks in plates and contains small scales of mica.

Clay No. 4 (bottom layer) is compact, of pearl gray color, uneven fracture, and containing small scales of mica."

The works have a capacity of 16,000 bricks a day, running full time, and the material shipped from them bears a deservedly high name.

These clays rest almost directly upon the seral conglomerate (XII) and are therefore at the bottom of the lower productive coal measures. The clays run regularly, varying of course in thickness, but keeping their general character and average size with sufficient constancy.

With the average dip to the north-west found at this point the clay deposit sinks steadily down and passes beneath water level of the Moshannon creek. No fire-clay is worked of those underlying the different coal beds of the first basin.

Fig. 61.

The next point where fire-clay is opened and worked is on the land of the Harrisburg Fire Brick Company, about two and a half to three miles west of Blue Ball station, on the Tyrone and Clearfield railroad, and 400 feet (by barometer) above the railroad level. The opening shows (Fig. 61):

Surface and clay, - - - - - -	2′	
Impure fire-clay, - - - - - -	7	
Hard fire-clay, - - - - - - -	4	10″
Soft fire-clay underlying.		
Massive white sandstone.		

Where exposed in a shaft, north-east of the open work, the clays show:

Surface and loose stuff, - - - - -	6′	
Coal, - - - - - - - - -	0	2″
Soft fire-clay, - - - - - - - -	6	
Hard fire-clay, - - - - - - -	7	

Another shaft gave:

Surface and loose stuff, - - - -	6	
Coal smut, - - - - - - - -	0	2
Fire-clay, - - - - - - - -	6	
Dark fire-clay, - - - - - - -	1	6
Soft fire-clay, - - - - - - - -	0	4
Hard fire-clay, - - - - - - -	3	
Brown sandstone, - - - - - - -	0	2
Massive sandstone, hard, white, in bottom.		

These clays are in three layers, called respectively, the upper layer, or "shell clay;" the middle layer, or "block clay," called the best of the three; the lower layer, or "flag clay."

Average specimens of these three clay lavers, forwarded for analysis, yielded (S. A. Ford):

	Upper layer.	Middle.	Bottom.
Silica, - - - - -	42.700	43.350	44.550
Alumina, - - - - -	37.600	37.550	39.000
Protoxide of iron, - -	2.385	2.145	1.440
Titanic acid, - - -	2.500	2.825	1.700
Lime, - - - - - - -	.112	.084	.028
Magnesia, - - - - -	.270	.234	.072
Alkalies, - - - - -	.730	.235	.530
Water and organic matter, -	13.840	14.170	13.660
	100.137	100.593	100.980

The upper layer is hard, compact, and of a dark bluish gray color.

The middle layer is hard, compact, of a dark pearl gray color, with conchoidal fracture.

The lower layer is hard, compact, of a light pearl gray color, with conchoidal fracture.

The mine from which the above specimens were taken is extensively worked by the company. They have no works at Blue Ball, but forward the clay raw to the Harrisburg Fire Brick Works, where it is manufactured into bricks. The works have a capacity of about 1,500,000 bricks per annum; easily increasable to double that amount.

The bricks are used for heating and puddling furnaces and for blast furnace linings; chiefly in the Schuylkill, Susquehanna and Cumberland valleys.

Moreover, the raw clay is shipped to Pittsburg to the Fire Brick Works there; and is used for making pots for the glass works. It is also shipped east, though not extensively, to Fire Brick Works.

The analysis of this clay shows it superior to the other Clearfield county clays examined; though the analysis of small specimens must always afford an only partially accurate comparison.

The bricks, however, bear a very high reputation, and the clay is in demand.

It will be noted that these clays carry an average two and a half per cent. of titanic acid, one of the Clearfield clays showing one-half per cent. of titanic acid, and the others none. It is an interesting question how far the superior heat resisting qualities of fire bricks are affected by the presence in quantity of this rather unusual constituent.

These clays in their floor, cover, character and size, resemble strongly the Sandy Ridge fire-clays, and give every evidence of being the same bed, altered but little in its passage under ground from the Sandy Ridge mine, on the crest of the Allegheny mountain, to this Blue Ball mine, where the clay is again raised high up and comes out to daylight near the summit of the First anticlinal sub-axis.

There is a bed of fire-clay struck in the wells at Wallaceton: and from the measures exposed in the vicinity it is very possible, and even probable, that it is the same bed as that exposed at Sandy Ridge and Blue Ball. But if so it seems to have lost, temporarily, both in size and character.

The Hope fire-clay works are at Woodland station, on the Tyrone and Clearfield Railroad, six miles east of Clearfield. The mines are opened on the south side of Roaring Run Brook, about 40 feet above the stream. Massive sandstone makes the country rock between the stream level and the floor of the mine. The hill rises 50 feet above, covered on the surface with sandstone lumps, usually of moderate size, without any pebble rock conglomerate.

The working face of clay exposed measured an average of

about five feet of hard, good looking clay, with softer or more impure fire-clay in roof and floor. While a part of this five foot clay occasionally deteriorated temporarily in character, yet the general average of the bed, both in size and quality, is sustained with much regularity.

Another drift, about 100 yards away, shows nearly the same thing: but with perhaps more of the inferior, and less of the valuable, clay showing in the working face.

An average specimen of the clay forwarded for analysis, yielded, (M'Creath):—

Silica,	46.250
Alumina,	37.500
Protoxide of iron,	1.935
Lime,	.168
Magnesia,	.126
Alkalies,	1.115
Water and organic matter,	13.540
	100.634

The clay is hard, compact, of a pearl gray color and somewhat slaty structure.

The Woodland Fire-clay works are on the Tyrone and Clearfield Railroad, about three-fourths of a mile west of Woodland station. The mine is opened on the north side of Roaring Run Brook, and shows 4 to 5 feet of good hard clay in places, but varying both much and rapidly, the workable layer in some places being pinched down very small. A small 3 or 4 inch coal overlies the workable clay layer, and on top of the coal there come in several feet of darker and impure fire-clay.

An average specimen forwarded for analysis yielded (S. A. Ford):

Silica,	45.450
Alumina,	36.125
Protoxide of iron,	2.275
Lime,	.168
Magnesia,	.342
Alkalies ,	1.290
Water and organic matter,	13.730
	99.380

The clay is hard, compact, of pearl gray color and slaty structure.

These openings last mentioned (Hope and Woodland) are apparently on the same bed; and there is every probability that this bed is the same as the one worked at Sandy Ridge and Blue Ball. It may be said that this fire-clay deposit, resting on top of the seral conglomerate and below the coals of the Lower Productive coal measures, is widely distributed and well known in Pennsylvania.

A fire-clay has been recently opened about three-fourths of a mile north-west of the Woodland station, on the Tyrone and Clearfield railroad. It is the same fire-clay as that opened, and already described, at the Hope and Woodland Fire-clay mines; but the three inch coal seam which is present at those openings is entirely wanting in this trial shaft.

About one-half mile south-east of the Hope mine, the bed has been opened again: and here the bed is fossiliferous, and the coal has thickened out to 5 inches.

Two average specimens yielded on analysis, (M'Creath):

	Hard clay No. 1	Soft clay. No. 2.
Silica, - - - - - -	45.230	46.180
Alumina, - - - - -	38.030	36.880
Oxide of iron, - - -	1.980	2.250
Lime, - - - - - -	.163	.173
Magnesia, - - - - -	.237	.317
Alkalies, - - - - -	.830	2.760
Water, - - - - - -	13.605	11.580

Clay No. 1, is hard and compact, with slaty color.

Clay No. 2, is compact, of pearl gray color, comparatively soft.

At Barrett Station, on the Tyrone and Clearfield Railroad, a fire-clay was opened some years ago, 95 feet (by barometer) above the railroad level, but was never worked for shipment to market. The clay was apparently of uncertain quantity, and had fallen off in quality.

Fire-clay works have recently been established at the town of Clearfield. The mine opening is made in the hill side, east of the Railroad depot, at the north end of the town. The section showing on the crop where opened is a curious exaggeration of an ordinary fire-clay deposit. It is as follows, beginning at the top:—

Fig. 62.
Clearfield. 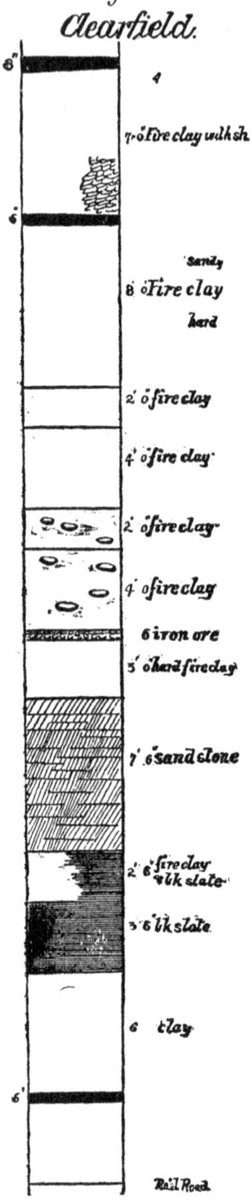

	ft.	in.
Coal,	6″ to 9″	
Impure fire-clay, with shales intermingled,	7′	0
Coal,	0	6
Hard fire-clay, sandy near the top,	8	0
Fire-clay,	2	0
Fire-clay,	4	0
Fire-clay, with nodular iron ore balls,	2	0
Fire-clay, some few iron ore lumps,	4	0
Carbonate iron ore balls,	0	6
Hard fire-clay, impure,	3	0
Sandstone,	7	6
Fire-clay and black slate,	2	6
Black slate, clayey,	3	6
Fire-clay,	6	0
Coal, Peacock,	0	6

Where the clay was opened in the mine for working at the time of the examination (July, 1874,) it showed:—

	ft.	in.
Top clay and sandstone,	4′	0′
Smooth clay,	2	6
Coarse plastic clay,	2	6
Coarse plastic clay,	1	6
Hard clay, used for terra cotta ware,	4	1
Clay,	2	3
Carbonate iron ore balls in clay.		

The bottom clay layer of this section shows great irregularity, making a waving floor, the layers above, however, being more smooth and even.

The different clay layers showed much diversity in character, some being quite sandy, while interleaved layers were evidently of excellent quality. The prevailing character of the clay, however, was over silicious, as the analyses given below plainly show.

Since the examination the valuable layers of this clay mass have apparently either diminished in size or retrogaded in quality, as the mine is now abandoned and clay for the works is brought from near Woodland Station, on the Tyrone and Clearfield Railroad.

A full suite of specimens of this Clearfield fire-clay, forwarded to the Laboratory of the Survey in Harrisburg, yielded on analysis, (specimens 1 and 2 by A. S. M'Creath, and specimens 3, 4, 5, 6, 7 and 8 by S. A. Ford):

	1.	2.	3.	4.	5.	6.	7.	8.
Silica	60.130	64.850	50.150	67.950	57.875	53.560	61.000	51.360
Alumina	25.710	23.770	35.600	20.150	27.005	28.820	25.800	31.250
Protoxide of Iron	2.371	1.218	.845	1.980	2.549	2.243	2.347	1.936
Bisulphide of Iron	.067	.032	......	......	.033	.135	.064	.748
Titanic acid	......	......	......	......	......	......	......	.500
Lime	.117	.190	.112	.084	.112	.431	trace.	.061
Magnesia	.663	.122	.160	.216	.465	.605	.530	.260
Alkalies	3.490	.345	.070	2.045	3.170	1.800	2.800	.035
Sulphuric acid	.191	.280	......	......	......	.869	.379	.381
Water	7.280	9.560	13.610	6.580	8.305	11,406	7.792	12.832
	100.019	100.367	100.547	99.005	99.514	99.869	100.712	99.363

These specimens were selected by the superintendent of the works from the fire-clay bed thus:

No. 1, ...	2′ thick
No. 2, ...	2 "
No. 3, ...	4 "
No. 4, ...	2 "
No. 5, ...	2 "
No. 6, ...	2 '
No. 7	4 "
No. 8	2 "
Iron ore balls in bottom.	

No. 1, is hard, compact, and of a slaty color.

No. 2, is hard, compact, and of a slaty color.

No. 3, is hard, compact, and of a dark olive color, fracture conchoidal, and structure slightly laminated.

No. 4, is hard, compact, and of a dark gray color.

No. 5, is hard, compact, unctuous, with gray color and slaty structure.

No. 6, is hard, compact and slaty, and of a slightly bluish color.

No. 7, is hard, brittle, unctuous and of a gray color.

No. 8, is hard, compact, of a light pearl gray color, with conchoidal fracture.

Tests were made in the summer of 1874, at the Clearfield fire-clay works of the standing up power of several well known fire bricks. The bricks tested were the Mount Savage, Dunkirk, (Scotch,) Woodland, Clearfield county, and two bricks made from the best interleaved layers of clay in the mine of the Clearfield fire-brick works. The temperature was that of molten steel. The Mount Savage brick, as might be expected from its high reputation, bore the test, but gave some signs of yielding.

The Dunkirk brick stood the test admirably, giving but slight signs of yielding, and the Clearfield bricks bore the test about equally well.

The fire-brick from the Woodland works was of good quality, but showed signs of yielding.

The crown of the kiln, in which the experiment was made, composed of Clearfield bricks, was intact at the end of the test, showing great standing up power.

It will be remembered that two points of great swelling in size of fire-clay beds on Clearfield creek have already been noted, though the quality in neither case seemed to be of the first order. It is difficult to locate exactly in its geological position this mass of fire-clay at Clearfield. From the record of the old oil boring, already given, in which only one small six inch coal was found under this clay, it would seem to be safe to assign to it the same geological horizon at the Sandy Ridge and Blue Ball fire-clays. But the measures of the Second Coal Basin, here exposed at Clearfield, are very difficult of identification. The absence of massive sandstone, the numerous and invariably moderate sized coals, the great development of clays and clay slates, render it difficult to bring the measures exposed here in harmony with those showing in the vertical sections made at Osceola or Snow Shoe, in the Eastern sub-division of the First Coal Basin, or at Karthaus, in the Western sub-division of the same basin.

This clay deposit is continuous for a distance, but was never observed in any other place in so exaggerated a condition as where showing at the Clearfield brick works.

In order to facilitate a rapid comparison of these fire-clays with some well known and valuable clays, the analyses which have been given in describing the fire-clay mines are grouped below into tabular form, taken from the report of Mr. M'Creath, Chemical Assistant to the Geological Survey of Pennsylvania; and below that a table of analyses of some of the best English fire-clays, the analyses being made by Mr. Abel, F. R. S., Chemist to the English War Department; the table of English fire-clays, as it stands, being taken from a valuable paper on fire-bricks by Lieut. Grover, of the Royal Engineer Corps. It is not intended to discuss the question of fire-bricks, but only to afford opportunity for comparison, and to state some few conclusions, Lieut. Grover's paper being freely drawn upon, verbatim.

With these tables there are also given two analyses of good clays from New Jersey, taken from the New Jersey Geological Report of 1868.

The Survey has necessarily confined itself in this report to a statement of the analyses of the fire-clays examined, there having been neither time nor opportunity for a more elaborate examination. But during the present year it is proposed to carry out a series of practical tests of the standing up power of these clays given above, together with other valuable clays in the State, testing at the same time for purposes of comparison, well known clays from other States and from England. These results, conveniently tabulated, must prove of value.

Analyses of Clearfield Fire-clays.

		Silica.......	Alumina	Protoxide of Iron.......	Titanic Acid,	Bisulphide of Iron.......	Lime.......	Magnesia....	Sulphuric Acid.......	Alkalies.....	Water, &c..	Total.......
Sandy Ridge....	1.	44,950	37.750	2.700	....	......	.302	.216	.075	.985	13.050	100.028
	2.	45.650	34.730	3.546	....	......	.112	.619	.165	5.750	9.650	100.222
	3.	45.820	35.950	3.330	....	......	.112	.573	......	4.130	10.130	100.045
	4.	74.950	15.940	1.899	....	......	.106	.407	.050	1.756	4.885	99.993
Blue Ball.......	1.	42.700	37.600	2.385	2.500	......	.112	.270	......	.730	13.840	100.137
	2.	43.350	37.550	2.145	2.825	......	.084	.234	......	.235	14.170	100.593
	3.	44.550	39.000	1.440	1.700	......	.028	.072	.....	.530	13.660	100.980
Clearfield.......	1.	60.130	25.710	2.371	....	.067	.117	.663	.191	3.490	7.280	100.019
	2.	64.850	23.770	1.218	....	.032	.190	.122	.280	.345	9.560	100.367
	3.	50.150	35.600	.845	....	......	.112	.160	......	.070	13.610	100.547
	4.	67.950	20.150	1.980	....	......	.084	.216	......	2.045	6.580	99.005
	5.	57.875	27.005	2.549	....	.033	.112	.465	......	3.170	8.305	99.514
	6.	53.560	28.820	2.243	....	.135	.431	.605	.869	1.800	11.401	99.869
	7.	61.000	25.800	2.347	....	.064	Trace,	.530	.379	2.800	7.792	99.712
	8.	51.360	31.250	1.936	.509	.748	.061	.260	.381	.035	12.382	99.363
Woodland St'n,	1.	46.180	36.880	2.250	....	......	.173	.317	.009	2.760	11.580	100.149
	2.	45.230	38.030	1.980	....	......	.163	.237	.013	.830	13.605	100.088
	3.	46.250	37.500	1.935	....	......	.168	.126	......	1.115	13.540	100.634
	4.	45.450	36.125	2.275	....	......	.168	.342	......	1.290	13.730	99.380

Prof. George H. Cook, in his report on the Geology of New Jersey, (1868,) gives the following as analyses of two good specimens of New Jersey clays:

	White clay, near South Amboy.	White clay Trenton.
Silica, - - - - - - -	43.200	45.300
Alumina, - - - - - -	39.710	37.100

	White clay, near South Amboy.	White clay, Trenton.
Zirconia, - - - - - -	1.400	1.400
Potash, - - - - - -	.370	1.300
Lime, - - - - -	——	.170
Magnesia, - - - - - -	——	.220
Peroxide of Iron, - - -	.740	1.300
Water, - - - - - -	14.250	13.400
	99.670	100.190

English Fire-clays.

	Silica.	Alumina.	Peroxide of Iron.	Alkalies, waste, &c.
Stourbridge, - -	65.650	26.590	5.710	2.050
Do. - -	67.000	25.800	4.900	2.300
Do. - -	66.470	26.260	6.630	.640
Do. - -	58.480	35.780	3.020	2.720
Do. - -	63.400	31.700	3.000	1.900
New Castle, - -	59.800	27.300	6.900	6.000
Do. - -	63.500	27.600	6.400	2.500
Burton-on-Trent, -	56.630	35.310	2.990	5.170
Wortley, - - -	65.200	29.690	3.070	2.040
Poole, - - -	68.600	23.600	4.700	3.100
Plympton, - -	75.890	21.610	1.960	.540
Do. - -	76.700	20.100	1.700	1.500
Hedgerly, - -	84.650	8.850	4.250	2.250
Holytown, - -	59.480	31.450	6.900	2.170
Dinas, - - -	96.200	2.000	.280	1.700
Kilmarnock, - -	59.100	35.760	2.500	2.640
Glenboig, - -	62.500	34.000	2.700	.800

These bricks have been tested practically at the Royal Arsenal furnaces, and the results, in Lieut. Grover's judgment, justify the conclusion that the refractory values of fire-bricks vary inversely with the amount of iron contained in them, and as a general rule the presence of six per cent. of peroxide of iron, warrants the absolute rejection of a fire-brick. This component usually takes the form of little black specks or mottled particles which are embedded in the material, and can be plainly detected upon breaking the brick.

Clays proper are chemical compounds, occurring under different phases in numerous geological formations, and consisting of hydrated silicates of alumina, either alone or in combination with silicates of potash, soda, lime, magnesia, iron, manganese, &c.

The so called fire-clays owe their refractory properties to a variable absence—differing, that is to say, in different clays—of lime, oxide of iron, and the alkalies of magnesia, potassa and soda. In refractory bricks, formed of baked fire-clay, the silica may be considered as a passive ingredient, acting mechanically to prevent excessive contraction, whilst the alumina forms the cement which binds the particles together.

The essential qualities of a good fire-brick may be classified as follows: Infusibility, regularity of shape, uniformity of composition, facility for cutting, strength and cheapness.

Infusibility seems to forbid in the brick's composition so much as even five per cent. of peroxide of iron, or three per cent. of combined soda, lime, potassa and magnesia. Generally speaking, a fire-brick should contain either silica or alumina in excess, according as it is intended for exposure to extreme heat, or for a possible contact with metallic oxides, which would exert a chemical re-action, decomposing it and acting as a flux. Thus, in theory, the arches of a furnace should be built of silicious bricks; its sides, bridge and neck of aluminous bricks. Dr. Percy considers that to boast properly of the quality of infusibility, a fire-brick must well resist sudden and great extremes of heat; it must support considerable pressure at a high temperature without crumbling; it should not melt or soften in a sensible degree by exposure to intense heat long and uninterruptedly continued; and it should withstand, as far as practicable, the corrosive action of slags rich in protoxide of iron. He recommends as a test, that the fire-clay should be formed into small sharp-edged prisms, which, on being enclosed in a covered crucible and subjected to an extreme temperature in an air or blast furnace, would denote a very high degree of refractoriness if the edges remained sharp, an incipient fusion of the material if the edges are rounded, and a thoroughly inferior quality of the fire-clay if the prisms were melted down.

Strength is obviously necessary to enable the bricks to avoid breakage in transport, and to withstand the pressure and cross strains to which they will be subjected when built into the work. It is stated by Rankine that the resistance to crushing by a direct thrust, the bricks being set on edge in an hydraulic press, is per square inch in weak red bricks, from 550 to 800 pounds; in strong red bricks, 1,100 pounds; in fire bricks, 1,700 pounds. But experiments made at the Royal Arsenal upon isolated cubes of one and a half inch side, cut from fire-brick "soaps," and placed between small squares of sheet lead, gave the following results:

	Cracking weight, lbs. per sq. inch.	Crushing weight, lbs. per sq. inch.
Stourbridge,	1,478	2,400
Stourbridge,	1,156	2,245
New Castle,	889	1,512
Plympton,	1,689	2,666
Dinas,	1,123	1,288
Kilmarnock,	2,134	3,378
Glenboig,	1,067	1,556

And the average crushing weight of ordinary stock bricks was found to be from 666 pounds to 866 pounds per square inch. Hence, all fire-bricks known may be said to have a strength far in excess of that which would ever be required of them in actual work.

Bricks with a high percentage of silica, though having a high refractory power, should not be exposed to the action of slags rich in metallic oxides, or to the fumes from lead ores, or to proximity with alkaline substances generally.

CHAPTER XXI.

The Third Bituminous Coal Basin continued down Bennett's Branch Creek, into Elk County.

In the reports on the Reynoldsville Gas Coal Basin, (Chapter XXII to XXVI,) and on the Fifth Coal Basin along Red Bank creek, (Chapter XXVII,) the line of the Low Grade division of the Allegheny Valley railroad is included, excepting that part which lies along the Bennett's Branch creek between its mouth at Driftwood and its head waters where the railroad leaves it at the Summit tunnel.

From the mouth of the Bennett's Branch, on the Susquehanna river at Driftwood, the Low Grade railroad following the bank of the deep cutting stream, runs for 10 to 12 miles through the measures of IX, the Ponent of Rogers, the hill which rise 800 to 1,000 feet on either side of the creek being capped by the massive white sandstones of the Seral Conglomerate. It is not until this Seral Conglomerate has come down from the mountain top into the Third Coal Basin, at Benezette, formerly Winslow, in Elk county, that the railroad touches the productive coal measures.

The high barren table land of the second anticlinal axis between the Second and Third Coal basins was examined for the First Geological Survey, and there is but little to add now to the description then given.

"Between the Karthaus hills of coal measures, in the Second* Basin, and those of Bennett's Branch, in the Third, there passes a wide tract of elevated barren country, consisting of the coarse rocks of Seral Conglomerate, which are here lifted to the surface by a broad and flat anticlinal axis. This axis of elevation, dividing the two coal fields, crosses the East Branch of Sinnemahoning a few miles above its mouth; then, curving gently southward, it passes the main Sinnemahoning at the mouth of the Driftwood Branch, to range in a west south-west direction through the high land dividing the waters of Clearfield creek from those of Bennett's Branch. The belt of country immediately along this line presents few mineral features of importance. The Seral sandstone and Conglomerate cap the highest hills. Indications of iron ore occur in a few bog-ore springs,

* It has already been explained that the Karthaus Coal Basin was erroneously placed in the Second Coal Basin in Rogers' Final Report of 1858; Karthaus, however, really being the Western sub-division of the First Coal Basin.

The broad high table land between Karthaus and the Third Coal Basin on the Bennett's Branch, the country rock everywhere being the seral conglomerate, represents the effect of both the First, or Laurel Hill, and Second, or Chestnut Ridge, anticlinal axes, the synclinal basin between them amounting to almost nothing and taking in on these hill tops, little or none of the productive coal measures.

visible near the Sinnemahoning; but the great height and steepness of the hills, at the very summits of which it must lie, would render the ore, even if abundant, difficult of access.

The synclinal axis marking the middle of the Third Coal Basin crosses the East Branch of the Sinnemahoning about 10 miles above its mouth, and the Driftwood Branch at the same distance from its outlet, following the course of the stream. On the west side of the Driftwood Branch one bed of coal has been discovered and opened near the level of the highest lands. A body of productive coal measures may possibly exist between this and Bennett's Branch, though from the obvious shallowness of the basin here, the existence of such appears improbable. The country is excessively wild and rugged. The apparent extension of the coal measures for five or six miles north-west of Bennett's Branch, up Hick's run and Trout run, is a matter which claims some attention on the part of proprietors of lands in this quarter." (Rogers' Final Report on Pennsylvania.)

At Benezette, in the Third Coal Basin, the measures have been somewhat developed. Mr. Young reports the following section as showing on the high lands north of the village, the top of the Seral Conglomerate being 280 feet above the Bennett's Branch creek. The lowest coal bed of the section was not seen. The measures show, (Fig. 140,) as follows:—

Fig. 140.

Hill top.		
Concealed measures, ...	10	
Coal, ...	2	4''
Concealed measures, ...	15	
Coal, ...	3	
Concealed measures, ...	10	
Coal, ...	2	6
Concealed measures, ...	10	
Coal, ...	1	6
Slates, ...	20	
Seral Conglomerate, (XII) . .	50 or more.	

The three foot bed of the above section is opened and worked. It looks very much as though these four small beds represent the one coal bed A; the bed partings having temporarily thickened out from inches to feet.

The fire-clay deposit on top of the Seral Conglomerate has not been opened at these mines.

Mr. Petriken, who owns this colliery, forwarded two specimens of his coal to Mr. M'Creath, at Harrisburg, for examination. The specimens were: No. 1 from the three foot bed; No. 2 from the two foot eight inch (top) bed.

These specimens yielded on analysis (M'Creath):

	No. 1.	No. 2.
"Water, - - - -	2.460	2.920
Volatile matter, - -	30.470	27.280
Fixed carbon, - -	62.227	59.725
Sulphur, - - -	.823	.705
Ash, - - - -	4.020	9.370
	100.000	100.000

No. 1, Coke per cent., 67.07. Color of Ash, cream.

The coal has a shining lustre, is columnar, friable, and containing iron pyrites.

No. 2, Coke per cent., 69.80. Color of Ash, brown.

The coal has a dull lustre, is somewhat slaty, and containing only a little pyrites."

These coals show quite free from sulphur, and one of them a little too ashy: but they should yield a good steam coal.

From a point about half a mile above Benezette, up to Trout Run, a distance of about two and a-half miles, the fire-clay overlying the seral conglomerate is in sufficient quantity and of such quality as to make its existence a matter of considerable importance to the region.

Mr. Young, who examined it, reports it as lying from 300 to 400 feet above the creek, and as showing from three feet to seven feet of hard fire-clay, sometimes overlaid by a foot or two of argillaceous sandstone, topped by a little soft clay. At other places the soil rests immediately upon the clay. It is at present obtained by stripping, and is worked to a moderate extent.

A specimen of fire-clay from the Jones mine yielded, on analysis, (S. A. Ford):—

Silica, - - - - - - -	44.450
Alumina, - - - - - -	38.945
Protoxide of Iron, - - - -	2.135
Lime, - - - - - - -	.173

Magnesia,	.155
Alkalies,	.760
Sulphuric acid,	Trace.
Water and organic matter,	13.287
	99.905

The clay is hard, compact, with conchoidal fracture and pearl gray color.

A specimen of fire-clay from the mine of E. Fletcher and Brother, two miles west of Benezette, yielded, on analysis, (S. A. Ford):—

Silica,	44.045
Alumina,	39.445
Protoxide of Iron,	.940
Lime,	.075
Magnesia,	.115
Alkalies,	.720
Sulphuric acid,	Trace.
Water and organic matter,	14.138
	99.478

The clay is very hard, compact, and of a light gray color.

The Benezette clays are of excellent character, as the above analyses show; and the similarity in the analyses, the mines being one and a-half miles apart, speaks well for the regularity of the clay deposit. There are no fire-clay works at Benezette as yet, but the raw clay is shipped for very considerable distances, much of it to Pittsburg, and gives entire satisfaction. The clay is in quantity, cheaply accessible, its average character is good, and fuel for burning is in the same hill side. The fire-brick trade, therefore, of the region, has a bright outlook in the future.

The Seral Conglomerate is high on the hills at this point, but at Caledonia some three or four miles to the south-west it has come well down to the stream. As the fall of the stream is slight this rise is due chiefly to the sharp rise of the basin to the northward, and renders it probable that a very small amount of Productive Coal Measures will be found still further to the north-east, on Trout run and Hicks' run.

In the gray slaty Vespertine Sandstones there occurs a bed

of inferior sandy limestone about four feet thick. This contains many fossil shells and other marine remains, including those of one or two species of fishes. Associated with this limestone we find many little yellowish balls of excellent iron ore. Between these and the seral rocks we meet with indications of the bed of iron ore everywhere so prevalent in the umbral red shales, which formation itself seems to have thinned out. Up a little run, to be seen one and a-half miles below Caledonia, we find a deposit of the soft bog-ore, derived, as usual, from springs flowing out of this part of the strata. We discovered a fine natural exposure of the bed of hard ore about one and a-half miles above Caledonia, on the edge of Bennett's Branch. At the latter place the current has cut away the loose rock and covering soil, leaving the ore in view under the overhanging seral sandstone, the bottom of which is within eight feet of the water. Immediately beneath the sandstone lies one foot of black shale, and under this three feet of brown shale, containing scattered nodules of ore, underlaid by a solid bed of ore, forming, with a very little shale, a bed between three and four feet thick. This overlies a band of fire-clay. A recent slide of earth and loose rock, somewhat concealing the layer, prevented our ascertaining its thickness with perfect precision. Rogers' Final Report.

The Seral Conglomerate, the exact thickness of which was not satisfactorily determined, varies from a moderately coarse sandstone to a rough quartzose conglomerate, in which the pebbles are occasionally larger than a turkey's egg. Large masses lie loosely scattered in great abundance; they present blocks ot all sizes and every degree of relative fineness of texture, and admit of being easily split into fragments adapted to almost any building purpose.

The following vertical section (Fig. 141) represents the measures exposed in this Third Coal Basin along the Bennett's Branch. Some of the beds were found exposed on Warner's run, and other thin layers were detected on Mead's run, lying a mile to the east; from these various data the somewhat detailed section of the rocks has been compiled. The more complete portion of the section terminates at the great bed of sandstone, 100 feet thick: but we have introduced several of the overlying beds in the order in which they were observed ascending the high lands north of the sources of the above streams. The section shows:

Fig. 141.

Bennett's Branch.

Stratum	Ft.	In.
Hill top.		
White sandstone, coarse, approaching a conglomerate, ...	50′	
Coal, indicated by a bench.		
Yellowish sandstone, ...	40	
(A bed of limestone either above or below this stratum.)		
Brown shale, thickness unknown.		
Coal, (opened at Bockaway's, on the turnpike,)	2	6″
Fire-clay, ...	0	6
Sandstone, gray, slaty, with some interstratified shale, ...	100	
Black and dark shales, containing kidney-form iron ore, ...	20	
Coal, (upper 6″ slaty,)	3	
Fire-clay, ...	1	6
Fire-clay, blue, good quality, upper part containing many iron nodules, ...	8	
Ore stratum, with fire-clay, (in all 3′ 6″) comprising oolite iron ore, 3″; blue clay, 1′; compact shale, blue and yellow, resembling externally a limestone, its upper part containing oolitic iron ore, ...	8	
Olive shale, containing shell ore, . . .	8	
Sandstone stratum, ...	1	
Shale, with abundant ore, lower part massive and silicious, ...	45	
Olive shale, ...	15	
Coal, ...	1	6
Blue slaty sandstone, ...	5	
Limestone, ...	0	10
Clay, ...	1	
Limestone, ...	3	
Fire-clay, ...	1	
Dark shale, (with ore in its lowermost 4′,)	10	
Olive shale, ...	2	
Coal, inferior quality, ...	1	6
Black shale, ...	3	
Coal, good, ...	5	
Black shale, ...	2	
Sandstone, ...	10	
Black shale, ...	4	
Sandstone and olive shales, with nodules of iron ore, ... , ...	30	
Limestone, ...	4	
Sandstone, ...	10	
Blue shale, ...	6	

(Rogers' Final Report.)

Coal is now opened and worked in the vicinity of Caledonia. Mr. Young constructed the following vertical section of the measures exposed on Goff's (formerly Pearsall) run, one mile north of Caledonia. The measures show (Fig. 142) as follows:

Fig. 142.

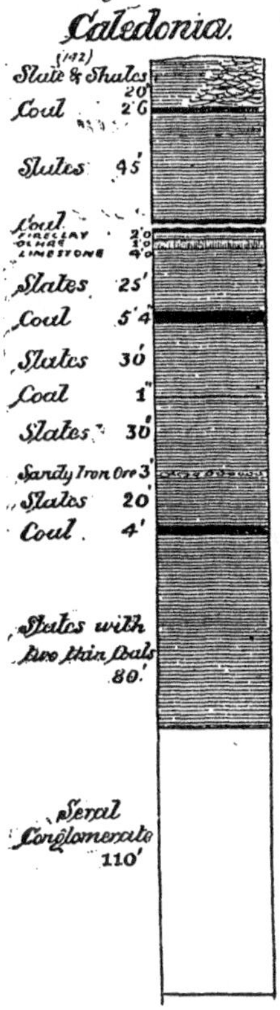

Caledonia.

Hill top.		
Slates and shales,	20′	
Coal,	2	6″
Slates,	45	
Coal, thickness not seen.		
Fire-clay,	2	
Ochre,	1	
Limestone	4	
Slates,	25	
Coal,	5	4
Slate,	30	
Coal,	0	1
Slate,	30	
Calcareous sandy iron ore,	3	
Slates,	20	
Coal,	4	
Slates, with two thin coals,	80	
Seral Conglomerate,	110	

Mr. Petriken has opened and is working the lower bed, the four foot bed. The coal yields on an analysis (M'Creath):

"Water,	1.770
Volatile matter,	32.170
Fixed carbon,	59.323
Sulphur,	2.067
Ash,	4.670
	100.000

Coke, per cent., 66.06. Color of Ash, reddish brown.

The coal has a dull lustre, is very friable, coated with silt, and containing considerable pyrites.

The large bed, 5 feet 4 inches thick, is worked at Goff's mine. The coal mines out well, and shows clean and bright with but little pyrites visible.

At the time of the examination for the First Geological Survey of Pennsylvania, Mr. Warner was working the "five foot" bed and the "three foot" bed. Analyses of these coals are found in Rogers' Final Report, as follows:

	5 ft. bed.	3 ft. bed.
Volatile matters, - - -	37.000	38.200
Coke, - - - - - -	63.000	61.800
Ash, - - - - - -	8.500	7.200

The iron ore lies in large slabs, three or four of them in the bed. It is calcareous and silicious.

These measures exposed on Goff's run, underlie a broad extent of territory at this place, and the beds are of workable size and the region well situated for a market. The show of iron ore is excellent, and now that the railroad renders transportation possible, the outcrop of these iron ores will undoubtedly be opened up. There is every reason to believe that regular and valuable deposits can be obtained on the outcrops given in the vertical sections above.

About one-half mile east of Caledonia, the following vertical section was made:

Hill top.	
Shales and thin sandstones, .	20′
Coal smut.	
Concealed measures, . . .	30
Coal bench.	
Slate and thin sandstone, . .	60
Coaly slate.	
Slate, ...	40
Coal smut, ...	2 or more.
Concealed measures, . . .	70
Seral Conglomerate, . . .	110

None of these benches or coal outcrops were opened up.

On the hill just north-west of the village, (Caledonia) the vertical section shows:

Coal, ...	2′ or more.
Shales and slates, ...	95
Coal, ...	2 or more.
Slates and shales, . . .	50
Seral Conglomerate, . . .	110

The old drifts on these two beds are now fallen shut. They are said to have averaged when worked, from 20 to 30 inches of coal.

Specimens of the so-called iron ore from Pearsall's (Goff's) run, one mile north of Caledonia, proved to be simply bluish limestones. Analysis of two specimens yielded, (S. A. Ford):

	1.	2.
Sulphur, - - - - -	.118	Trace.
Phosphorus, - - - -	.072	.031

Carbonate of Lime, - -	66.912	36.764
Carbonate of magnesia, -	9.836	2.011
Insoluble residue, - - -	16.130	53.330

No. 1 is a hard, compact limestone, minutely crystalline, with conchoidal fracture, of bluish grav color, and showing crystals of Pyrites.

No. 2 is a hard, compact limestone, exceedingly silicious, and of a bluish gray color.

This Third Coal Basin extends southwestward up the Bennett's Branch creek, the benches of the same coals and iron ores exposed at Caledonia showing in the hills along the sides of the creek. But until within the past year the country has been far removed from any cheap communication with a market: and since that time the business depression has checked materially the development of iron and coal properties. There is but little opening, therefore, found on the Bennett's Branch, above Caledonia.

Near Tyler's station, on the Low Grade Railroad, a coal is opened and worked to a small extent. It shows:—

Shales overlying.	
Carbonated clay slate roof, - - - - -	3′ 0″ or more
Coal, with small slate partings, not persistent.	3 0
Soft fire-clay floor.	

The coal in mining out shows sulphur balls in the irregular slate.

A specimen of this Tyler mine coal yielded, on analysis, (M'Creath):—

"Water, - - - - - - - -	0.940
Volatile matter, - - - - - -	31.060
Fixed carbon, - - - - - -	61.563
Sulphur, - - - - - -	1.487
Ash, - - - - - - -	4.950
	100.000

Coke, per cent., 68.00. Color of Ash, gray, with red tinge.

The coal is bright, friable, containing slate, charcoal and Iron Pyrites. The coke does not swell much in coking, and forms a hard, compact coke with metallic lustre."

Fig. 143.

Tylers.

A vertical section of the measures exposed at Tyler's, on the north side of the Sandy Lick creek, shows, (Fig. 143,) as follows:—

Hill top.	
Sandstones and slates, ...	45′
Bench.	
Thin sandstones and shales, . .	45
Bench.	
Slates and thin sandstones, . .	50
Bench.	
Sandy gray slates, ...	35
Bench.	
Shales and gray slates, ...	99
Coal bench, opened, ...	1 6″
Shales, some thin sandstones, .	83
Coal, Tyler's, ...	3 or more.
Concealed measures, . . .	30
Creek level.	

This hill should contain the five foot bed of Caledonia, and as the hills rise about 450 feet above the water on the south side of the Bennett's Branch, these measures pass into and through all the hills here and cover an extensive area with productive measures.

A four foot coal is reported by Mr. Tyler as once opened up two miles west of Tyler's station.

This coal was not open at the time of the examination, and could not be measured. The region between Tyler's and the head waters of Bennett's Branch, at the Summit Tunnel, consisting mainly of the same measures exposed in the section at Tylers, is but little developed.

CHAPTER XXII.

The Fourth or Reynoldsville Gas Coal Basin.

The sketch map (Plate 1) shows the position of the Third and Fourth anticlinal axes, which bound this Fourth Coal Basin.

The Third anticlinal axis continues as a very decided mountain all through Elk county and the extreme north-western point of Clearfield county, bringing up on the crest of the "Elk" or "Boon Mountain," the lower coal measures, (XI), below the Seral Conglomerate, on the centre of the axis. But the anticlinal is sinking rapidly to the south from this point, and where it crosses the Sandy Lick creek, not far from the Clearfield-Jefferson county line, and about five miles north 55° east from Reynoldsville, not only the measures of XI, but also the massive Seral Conglomerate, (XII), have passed beneath water level, and the gray slates and thin sandstones of the lower part of the Lower Productive Coal Measures arch gently over its broadened back. From here its course is south-west for many miles, passing about two miles north-west of Punxatawney and crossing Mahoning creek; south of which it takes up the "barren measures" of Rogers (the measures between the Mahoning sandstone and the Pittsburg Coal Bed) upon its back, and continues on far to the south-west, with its arch much flattened.

It may be noted here that this Third anticlinal axis runs about north 55° east from the Sandy Lick creek, along the crest of Boon mountain, while its general course south-west from Sandy Lick creek to Punxatawney, is about south 40° west, (north 40° east.) These axes apparently lie *en echelon*, in the same way as the First anticlinal sub-axis in the country west of Osceola, in Clearfield county, as already described; and while the country in the vicinity of the county line on Sandy Lick creek is not favorable for making exact determination of the position of the subordinate axes, the axis running north-east from the Sandy Lick (the Boon mountain axis) lies somewhat east of the line of the axis coming up from Punxatawney.

The Fourth anticlinal axis which makes the west side of this basin, is seen to cross the Little Toby creek, about three miles north-west of Brockwayville, Jefferson county. Here it is very marked, the massive seral conglomerate (XII) at the base of the Lower Productive Coal Measures, making the bed of the creek, and the country rock of the hill sides. Thence the axis runs south-west, Mill creek following its general line, and crosses the Sandy Lick creek at or near Port Barnet, two miles east of Brookville, Jefferson county. From here its course is steadily to the south-west, until it crosses the Mahoning creek and, like the Third anticlinal axis, receives upon its back the "Barren Measures."

The exposure of Seral Conglomerate along this Fourth anticlinal axis illustrates the rapid and decided changes in the character of this rock. The whole valley of the Little Toby creek is filled with great masses of true conglomerate rock, made up of white quartz pebbles, rounded, ranging in size from an egg to an olive or pea. The rock is almost wholly made up of these pebbles. And yet twelve miles away, on the Sandy Lick creek, very few layers of true conglomerate rock occur, the mass of this Seral Conglomerate being made up of massive fine grained sandstones, usually white, and for long distances without a trace of conglomerate.

The bed of the Sandy Lick creek, where the Fourth anticlinal axis crosses it, is made up of the sandstones of X, the measures of XI being here thin and unimportant.

While the Third and Fourth anticlinal axes mark the eastern and western sides of the Fourth Coal Basin, the geology of the basin is complicated by a subordinate but decided anticlinal axis, which crosses the Sandy Lick creek about two miles west of Reynoldsville. This axis brings up the measures of XII, the Seral Conglomerate, to the railroad and even above it; and from the centre of this subordinate axis westward to the Fourth anticlinal axis at Port Barnet, and beyond, the Sandy Lick creek flows in the Seral Conglomerate and the rocks just underlying.

The effect of this subordinate anticlinal axis upon the value of the basin is very marked; confining the "Big Bed" to that part of the basin (in the vicinity of the Sandy Lick creek) which lies between the subordinate anticlinal and the Third anticlinal axis, and allowing to that part of the Fourth Basin which lies between the subordinate anticlinal axis and the Fourth anticlinal axis, only the measures between the Seral Conglomerate and the Big Bed, restricting the greater part of it to one or two coal beds, instead of the whole productive series between the Seral Conglomerate and the Mahoning sandstone, which is exposed in the eastern half of the Fourth Basin.

Within the limits marked by the Third and Fourth anticlinal axes, the whole of the Lower Productive Coal Measures, from the Seral Conglomerate to the Mahoning sandstone, are exposed. The "Lower Freeport" coal bed, the "Big Bed" of the region, is the one now worked, and gives chiefly the present prominence and value to the Fourth Coal Basin. The "Reynoldsville Gas Coal," as now shipped, comes from this bed.

The Reynoldsville Gas Coal Basin map, (Plate VII,) shows how the Third anticlinal axis, leaving the Sandy Lick creek near the county line, runs south-west, keeping east of the M'Creight, Strouse, Elble and P. Hawk coal mines, crossing Elk run two miles north-west of Punxatawney. The Fourth anticlinal axis on the west side, keeps west of the Norris, Uplinger, Brown, Ruth and Straighthoof coal mines. The general eastern and western edges of the Lower Freeport Coal Bed are shown by dotted lines, and the cross section indicates how this coal necessarily rises out to daylight, and is not found east and west of the points indicated.

The section also shows how the measures below the Lower Freeport Coal, and the Seral Conglomerate, come out to daylight on the sides of the basin and sink below water level at its centre.

The gradual and decided thickening of the measures between the Seral Conglomerate and the Mahoning sandstone, going westward, the thickening also of the "Barren Measures," and the thinning down of the Seral Conglomerate and the rocks below it, will be discussed at the close of this report; the facts, however, being largely embodied in the detailed sections of the measures exposed in the Fourth and Fifth Coal Basins, and the subject requiring little more than the simple re-statement and comparison of the sections.

The dotted lines of the Lower Freeport Coal Bed outcrop (Plate VII) only enclose a small area north of the Sandy Lick creek, at Pancoast station on the Low Grade division of the Allegheny Valley railroad. The great body of this coal in the basin, above water level, lies between the Sandy Lick and Mahoning creeks. The reason for this is obvious. The whole basin is sinking slowly but steadily to the southward. The Lower Freeport Coal Bed, which is 1,480 feet above tide water at Reynoldsville, passes beneath the Mahoning creek, 15 miles to the south-west, at 1,193 feet of tide water. This represents the average sinking for these 15 miles. Of course, this sinking is not perfectly even and regular at every point, but is, in this basin as in all others, disturbed by slight cross rolls, at varying angles with the line of the basin itself. These cross rolls are not usually very great, rarely throwing the measures more than 25 feet out of level.

The steady rising of the basin to the northward, and the sharp rise to the north-west, caused by the subordinate anticlinal axis, shoots the Lower Freeport coal into the air just north of Pancoast station; and the measures exposed from Sandy Lick creek north-eastward to the Little Toby creek, are the rocks below the Lower Freeport Coal Bed, the latter never coming into the hill tops.

The Fourth Coal Basin, therefore, naturally sub-divides itself into two parts:

1. The basin extending from Pancoast station, on the Low Grade railroad, south-westward to the Mahoning creek; in which the Lower Freeport (gas coal) coal is worked.

2. The basin extending from Pancoast station, north-east to the Little Toby creek at Brockwayville; in which the coal beds lying below the Freeport coals are worked.

CHAPTER XXIII.

Reynoldsville and Punxatawney Main Coal.

The measures exposed above water level in the centre of the eastern side of the Fourth Coal Basin, are as shown in the accompanying vertical section, (Fig. 63.) The section is compiled from exposures at Fuller's Hill, one mile east of Reynoldsville, with some few additions. It shows:—

Fig. 63.
Fuller's Hill

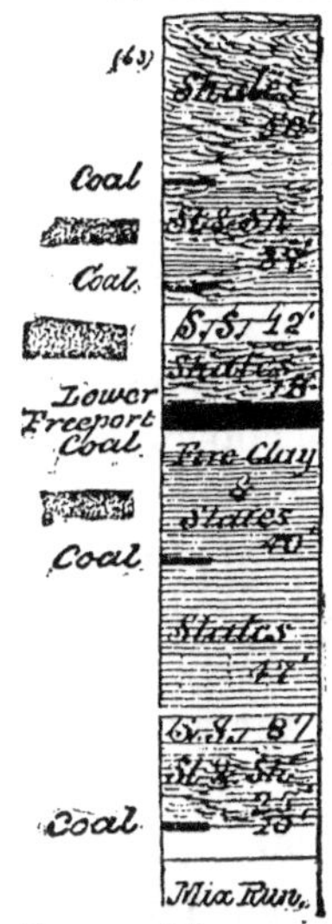

Hill top.	
Shales,	50'
Coal, not opened.	
Slates and shales,	39
Small coal?	
Sandstone,	12
Shales,	18
Coal, Lower Freeport,	7
Fire-clay,	5
Slates, grayish,	35
Coal outcrop.	
Gray slates, sandy,	47
Sandstone,	8
Thin gray slates and shales,	25
Coal? small.	
Concealed measures,	10
Level of Mix run.	

Of the coal beds in the above section, the Lower Freeport Coal is the "Big Bed," the "Gas Coal" bed of the region.

The coal beds below the Lower Freeport, will be described in detail, where they are fully opened and tested, and the bed above the Lower Freeport Coal, where it has been opened and worked in several places in a small way for domestic use.

The Lower Freeport Coal Bed is opened up and worked in many places, in some quite extensively for shipment to market; in many others in a small way to supply the local demand. The detailed description of these different mines shows that while the bed has considerable tendency to variation in size and character, yet as a whole the coal bed exists as a large and valuable deposit throughout the whole of this basin lying between the Sandy Lick and the Mahoning Creeks. The location and name of the mines worked in the basin, and the elevation above tide. will be found on the sketch map, (Plate VII.)

The Pancoast mine was opened on the north side of the Sandy Lick creek, one mile east of M'Ghee's mills. The main entry runs in north 15° west, and at 50 yards in measured as follows, (Fig. 64):

Fig. 64.
Pancoast 1.

Sandstone roof.

Coal,	1′ 3″ to 1′ 4″
Slate,	0 1 to 0 2
Coal,	5
Slate,	0 1 to 0 $1\frac{1}{2}$
Coal,	0 6

Limestone, clayey, bastard, in floor.

At about 60 yards in on the main entry, the sandstone roof gives place to a soft fire-clay, which comes down and cuts the coal to four feet, the floor remaining level. At 100 yards in the coal suddenly rises to nine feet in thickness, then pitches steeply to the east and north, and cuts down to nothing. A cross heading, driven off to the west 20° north, at 50 yards found the coal down to 15 inches. A heading was driven to the east and in 10 yards distance the coal was down to three feet and rapidly diminishing. The main entry was continued for 30 yards, through hard fire-clay and sandstone, and the mine was abandoned. The line of trouble to the coal seemed to be about north 65° west and south 65° east.

The trouble does not appear like a clean cut fault; but much more like a true horse-back. By that term is meant not merely a wedge of clay or rock, but the mass of clay and rock which represents now what was a stream, winding sluggishly through the low lands in which the coal vegetation grew. This stream filled with sand and loose stuff; and the horse-back therefore widens and narrows exactly in accordance with the irregular banks of the original creek. The line of trouble, moreover, follows no regular traceable line as in the case of a fault, but runs in any direction, governed only by the law which regulated the stream course.

The measures at this point (Pancoast Station) have been opened up on the outcrop to show the size of the coal beds underlying the Big Bed. The section shows, (Fig. 65)

Fig. 65.

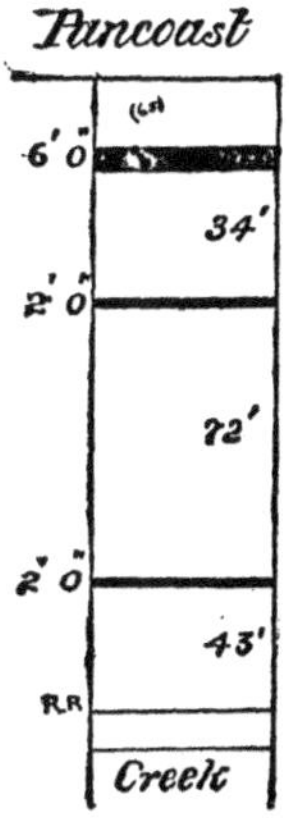

Pancoast

Hill top.		
Slates,	. . . . . .	60′ or more.
Coal, Lower Freeport,	. .	6
Slates (?)	. . . . . .	34
Coal,	. . . .	2
Slates and *concealed measures,*		72
Coal	. . . .	2
Concealed measures,	. .	43
Level of Sandy Lick creek.		

Where this Lower Freeport coal was opened in the troubled mine at Pancoast, it carried fine, black, hard, clean coal from the outcrop in to the trouble

The same coal bed was re-opened some 700 yards east of this abandoned mine, across a small ravine, near Pancoast Station, on the Low Grade railroad. The coal measured as follows, (Fig. 66):

Fig. 66.

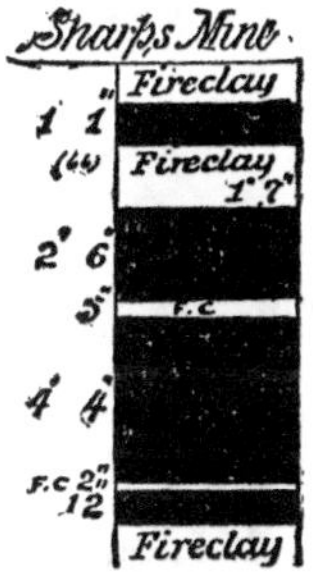

Sharps Mine.

Roof, fire-clay,	. . .		3′ or more
Coal, rotten and slaty,	. .		1 1″
Fire-clay,	. . .	1′ 6″ to	1 8
Coal,	. . . . . .		2 6
Fire-clay parting,	. .	0 4 to	0 6
Coal,	. . . .		4 4
Fire-clay,	. . . . .		0 2
Coal,	. . . .		1 0
Fire-clay floor, soft.			

At 140 yards in these measurements are changed, the fire-clay being replaced by slates, thus:

Coal in roof,	. . . .	2′ 6″ (?)
Slate and bone coal,	. .	0 4
Coal,	. . . .	4 4
Black slate,	. . . .	0 2
Coal,	. . . .	0 10
Fire-clay floor,	. . . .	2 0 or more.

The coal, in this opening, showed a coating of fine clay, especially along the butt.

These sections and descriptions are given in full especially to call attention to the remarkable and sudden changes which may be looked for in mining, even in a basin where the coal bed is running with much regularity on the average, as is the case with this Reynoldsville Gas Coal Bed. In this case, along a well defined bench, within a half mile, the floor, the coal and the roof show such changes as to make the bed almost unrecognizable.

An average sample of this Sharp mine coal forwarded to Mr. M'Creath for analysis, yielded, for the upper bench:—

"Water,	1.320
Volatile matter,	31.440
Fixed carbon,	62.578
Sulphur,	.892
Ash,.	3.770
	100.000

Coke, per cent., 67.24. Color of Ash, gray.

The coal is resinous, compact, iridescent, and contains little iron pyrites." It forms a compact coke, with silvery lustre.

An average sample of the lower bench yielded, on analysis, (M'Creath):—

"Water,	1.570
Volatile matter,	33.430
Fixed carbon,	61.285
Sulphur,	1.055
Ash,	2 660
	100.000

Coke, per cent., 65.00. Color of Ash, brown.

The coal is resinous, compact, with a coating of silt, and much Pyrites." It forms a coherent coke, with silvery lustre.

The above analyses show the coal to be of excellent character at this point.

The Pancoast and Sharp mines, as well as the Diamond and W. Reynold's mines, described hereafter, are on the property of the Central Land and Mining Company; to the President of which, Mr. Albert Pancoast, and to the Engineer, Mr. Smith, the survey is indebted for many courtesies.

The preliminary borings made on the south side of Sandy Lick creek, just opposite Pancoast station, to determine the most available position for opening this Lower Freeport coal, are reported as having been unsatisfactory. It is entirely possible and even probable, that this irregularity in the bed, shown by the borings, is connected with the troubled Pancoast mine region: the difficulty crossing the Sandy Lick creek in this direction.

The bed, however, has been opened upon the outcrop on the south side of the Sandy Lick creek, west of the point above named, and also opposite M'Ghee's mills. At both these places the outcrop openings give every promise of an average thickness and quality for the coal.

Near what is known as the "Old Clark Bank," two and a half miles north-east of Reynoldsville, there is a natural opening in a ravine, which shows as follows, (Fig. 67):

Fig. 67.
Clark.

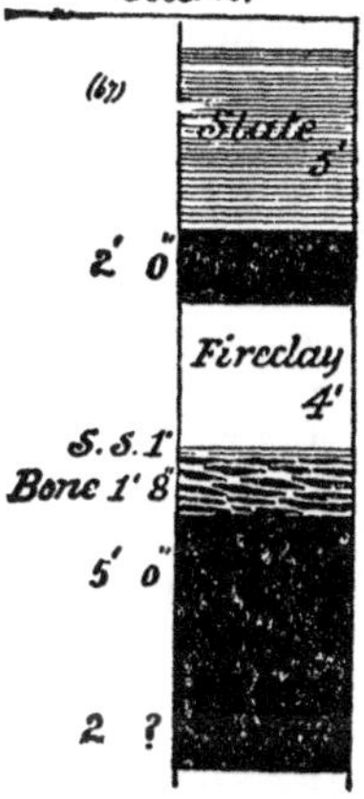

Slate on top, ...	5'	
Coal, ...	2	
Fire-clay, ...	4	
Sandstone, ...	0	1"
Bone coal, ...	1,	8
Coal, ...	5	
Coal in floor, (?)	2	

The coal is hard, black, shining, and shows very free from iron pyrites.

Sixty-seven feet above this coal is a bench of the upper coal.

With reference to the identification and naming of this bed above, the doubt is whether it should be called the Middle or Upper Freeport Coal.

In the sections of the First and Second Coal Basins, it has already been shown that there are three Freeport Beds, (instead of two,) an Upper, Middle and Lower Freeport Coal. The interval between the Lower and Middle coals is usually fifty feet or more; a limestone in many cases underlies the Middle coal, and an interval of fifty feet or more separates the Middle from the Upper coal. Moreover a section made in the Fifth Coal basin shows 102 feet again between the Lower and Upper Freeport beds, the Middle Freeport coming in between. It remains therefore, to decide whether the Lower Freeport coal, the small bed 35 feet above, and the bed 70 to 75 feet above, represent the three Freeport Beds, or whether the small coal 35 feet above, is only a rider, sporadic in character, and the Upper bed really the Middle Freeport coal. The lack of evidence of anything resembling a coal above the upper of these beds rendered it safer to call the three of them the Freeport Group. Of these, the Lower and Upper alone are persistent; the Middle coal only showing in a few places, and then apparently as a small bed.

The Lower Freeport Coal Bed shows itself plainly in a beautifully marked bench in all the ravines south of the Sandy Lick in this part of the basin, and is opened for extensive shipment by the Diamond Colliery, near Reynoldsville, their main entry running through the hill, (here very narrow,) from the outcrop on Pitch Pine run, to the outcrop on the south side of Sandy Lick creek. At this latter point, the coal is lowered down by an inclined plane on to the Low Grade railroad, about 135 feet below the outcrop.

The coal as measured in the Diamond mine showed (Fig. 68) as follows:

Fig. 68.

Roof, fire-clay,	15′ or more.
Coal,	2 1″
Fire-clay,	0 7
Coal,	1 11
Fire-clay,	3 4
Bony coal,	0 6
Coal,	6 1
Black slate,	0′ 2″ to 0 4
Fire-clay floor.	

Of the main bench of the coal, the six foot bench, the lower one and a half feet showed softer, friable coal; the upper four and a half feet of the bench being very hard and bright.

A specimen of the lower part of this "6 foot" bench, forwarded to Mr. M'Creath for analysis, yielded:

"Water,	1.120
Volatile matter,	33.860
Fixed carbon,	60.692
Sulphur,	1.278
Ash,	3.050
	100.000

Coke, per cent., 65.02. Color of Ash, gray, with reddish tinge. The coal has a dead lustre on outer surface; when broken against the layers it has a dull resinous lustre; with thin films of clay silt. When broken with the layers it has a bright, shining lustre. Contains soft mineral charcoal and iron pyrites in veins. It forms a coherent, compact coke, with metallic lustre.

A specimen of the upper part of this six foot bench yielded (M'Creath):

"Water,	0.980
Volatile matter,	30.300
Fixed carbon,	50.521
Sulphur,	2.429
Ash,	15.770
	100.000

Coke, per cent., 68.72. Color of Ash, gray, with red tinge.

The coal has a dull resinous lustre, contains charcoal and much iron pyrites. It forms a coherent, compact coke, with dull metallic lustre."

This specimen was taken from near the top of the six foot bench. It proved, what the appearance of the coal seemed to show, that the five feet of coal, starting from the bottom, represent the best of the coal bed.

An average specimen of the main 5 foot bench, the coal shipped to market, yielded on analysis (M'Creath):

Water,	.950
Volatile matter,	35.130
Fixed carbon,	59.304
Sulphur,	1.436
Ash,	3.180
	100.000

Coke per cent., 63.920. Color of Ash, dirty gray.

The coal has a dull resinous lustre generally, but with veins of bright crystalline coal running through the mass, is rather hard, and contains soft mineral charcoal. It forms a coherent coke, with a metallic lustre.

A specimen of the middle bench, (the 23 inch coal above the six foot bench) yielded, (M'Creath):—

"Water	1.190
Volatile matter,	32.810

Fixed carbon, - - - - - -	55.316
Sulphur, - - - - - - -	2.284
Ash, - - - - - - - -	8.400
	100.000

Coke, per cent., 66.00. Color of Ash, gray, with pink tinge.

The coal has a shining lustre, is compact, hard, and contains veins of pyrites."

This bench is not worked.

A specimen of the upper bench (the 25 inch coal) yielded, on analysis, (M'Creath):—

"Water, - - - - - - -	1.100
Volatile matter, - - - - -	29.990
Fixed carbon, - - - - - -	46.639
Sulphur, - - - - -	3.101
Ash, - - - - - - - -	19.170
	100.000

Coke, per cent., 68.91. Color of Ash, gray, with reddish tinge.

The coal has a dull lustre, is hard, compact, with veins of Pyrites and much slate."

The above analyses of these two upper benches speak for themselves, and condemn the coals.

A specimen of coke, roughly made in the open air at the Diamond colliery yielded, on analysis, (M'Creath):—

"Water, - - - - - - -	.500
Volatile matter, - - - - - -	1.150
Fixed carbon, - - - - - -	88.478
Sulphur, - - - - - - -	1.022
Ash, - - - - - - -	8.850
	100.000

Color of Ash, cream.

The coke has a shining lustre, is hard, compact, with pieces of slate through it."

This coke was made merely from the slack of the mine, without care to avoid the presence of slate, and is not equal to the coke which would be yielded by the solid coal of the mine. The coke is well spoken of where used.

An analysis of this coal for Phosphoric Acid (M'Creath) yielded only a trace. These analyses are labelled in the tables of Mr. M'Creath as the D. Reynolds mine.

This same Lower Freeport coal is opened and worked at Hoover's mine on the T. Reynold's property, one mile south of Reynoldsville. The coal measured in the mine at Hoover's, (Fig. 69,) as follows:—

Fig. 69.

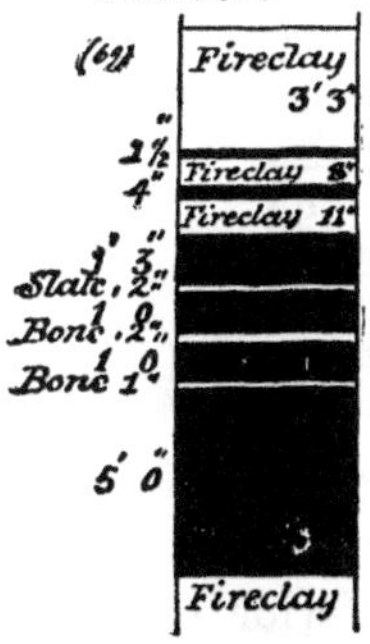

Fire-clay roof,		3′ 3″
Coal,		0 1½
Fire-clay,		0 8
Coal,		0 4
Fire-clay,		0 11
Coal,		1 3
Slate,		0 2
Coal,		1 0½
Bony coal,	0′ 2″ to	0 3
Coal,		1 0
Bony coal,	0 ½ to	0 1
Coal,		5 0
Fire-clay floor, hard.		

The bench of this coal bed which is now worked is the 6 foot coal (with the small parting at the top) at the bottom of the bed.

A specimen of the upper part of this bench forwarded for analysis, yielded, (M'Creath):

"Water,	0.960
Volatile matter,	32.680
Fixed carbon,	59.097
Sulphur,	1.063
Ash,	6.200
	100.000

Coke, per cent., 66.36. Color of Ash, gray, with red tinge. The coal has a dull lustre, is compact, hard, with iron pyrites in thin layers."

The middle part of the bench yielded on analysis, (M'Creath):

"Water,	1.100
Volatile matter,	30.800
Fixed carbon,	62.524
Sulphur,	.776
Ash,	4.800
	100.000

Coke, per cent., 68.10. Color of Ash, cream.

The coal has a dull, resinous lustre, is very hard, with charcoal and iron pyrites.'

The lower part of the bench yielded, on analysis, (M'Creath):

"Water, - - - - - -	1.100
Volatile matter, - - - - -	32.900
Fixed carbon, - - - - -	62.174
Sulphur, - - - - - -	.726
Ash, - - - - - - -	3.100
	100.000

Coke, per cent., 66.00. Color of Ash, cream.

The coal has a shining lustre, is compact, with veins of pyrites, charcoal and slate."

These complete analyses of the Diamond and Hoover mine coals agree in showing that the lower 5 feet of coal represent the best part of the bed; the upper benches being inferior to it. The coal is hard, mines out in lumps, and is well fitted to bear transportation.

In analysing the Hoover coal for Phosphoric Acid, Mr. M'-Creath found:

	Per ct. in Coal.	Per ct. in Ash.
Upper part of "6 foot" bench, - -	071	1.145
Middle part of "6 foot" bench, - -	.008	.166

A specimen of coke made in the open air, from the slack of the Hoover mine, the slack of course coming mainly from the bottom bench, yielded (M'Creath):

"Water, - - - - -	.780
Volatile matter, - - - - -	1.420
Fixed carbon, - - - - -	88.950
Sulphur, - - - - - -	.900
Ash, - - - - - - -	7.950
	100.000

Color of Ash, reddish gray

The coke is gray, compact, coherent, with lumps of slate, slightly iridescent, and comparatively soft."

The ash of the Hoover coal from the lower part of the six foot bench, yielded on analysis (M'Creath):

Silica, - - - - - -	1.220

Oxide of iron, - - - - - -	.420
Alumina, - - - - - - -	1.215
Lime, - - - - - - - -	.120
Magnesia, - - - -	.090
Phosphoric acid, - - -	.008
Percentage of ash in coal, - - -	3.100

East of the Hoover mine, on the same continuous well marked bench of the Lower Freeport coal, an old opening, now fallen shut, is said to have shown sandstone resting directly upon the coal.

An opening driven in on the coal a few hundred yards further east (Prescott's) shows:

Sandstone, ...	4′	0″	or more.
Crushed slate, ...	0	4	
Coal, ...	2	4	
Slate, ...	0	1	
Coal in floor.			

The mine is now fallen shut and was abandoned because, further in, the sandstone roof came down, cutting out the coal.

It is perfectly possible that the trouble already described at the Pancoast mine, the failure to find the coal satisfactorily on the south side of the Sandy Lick creek, opposite the Pancoast mine, and this cutting out of the coal at the Prescott mine, indicate the general direction of the horse-back trouble. There is no evidence of a fault of any size; for the coal is found, regular in size and character, to the north, south, east and west of this troubled opening, and keeping its level in this flat central part of the basin.

In many of the cases where sandstone rests directly upon coal, without the interposition of carbonated clays or slates, there is strong evidence that the sandstone is not the original roof. We are required to suppose that the vegetable matter was covered over completely, perhaps even for a sufficient length of time to allow the coal formation process to begin. Then a sharp local denudation, scouring out a deep trough, cutting away the whole or a part of the coal matter, and if only a part, depositing, rapidly, sand upon its top. Naturally, in such cases there is no law to govern the trouble except that which determined the direction of the eroding current. In many cases this partial replacement of slates by sandstone is found only over a limited area, and with the general run and character of the bed undisturbed.

The Lower Freeport coal is opened at the Woodward Reynolds' mine, one mile south of Reynoldsville station. The coal shows as follows (Fig. 70):

Fig. 70.

W. Reynolds.

Roof, clayey black slate,	4. or more.	
Black slate and bone coal,		
Coal,		
Slate and bone coal,	0	3'
Coal,	0	6
Slate,	0	3
Coal,	3	8
Black slate,	·0	1
Coal,	1	1
Slate,	0	1
Coal,	0	7
Fire-clay,	4	
Limestone,	4	
Shale.		

This mine is not worked for shipment to market; but is an old mine and has been considerably worked for local use.

A specimen of the coal forwarded to Mr. M'Creath for analysis, yielded:

"Water,	1.440
Volatile matter,	32.460
Fixed carbon,	63.011
Sulphur,	.639
Ash,	2.450
	100.000

Coke, per cent., 66.10. Color of Ash, cream.

The coal has a shining lustre, is compact, with a small amount of iron pyrites."

It is worthy of note that these four mines above described, yielding by analysis the same character of coal, vary in their roof and partings so markedly in the short distance of one mile. At the Diamond colliery, the roof, partings and floor are all of soft fire-clay; at Hoover's, a roof of hard fire-clay, and partings of black slate; at W. Reynolds', a roof of black clay slates, black slate partings, and limestone under the fire-clay floor; at Prescott's and the old T. Reynold's mine, a sandstone roof. If the identification of this bed were to depend on appearance alone, there would be ample ground for denying that these openings were all on the same bed; but the bench goes around the hill sides, perfectly marked, smooth and regular, from mine to mine, and the identification is complete.

The same coal bed is opened and worked at Shiesley's mine, one mile south south-east of Reynoldsville. The coal measured in the mine (Fig. 71) as follows:

Fig. 71.

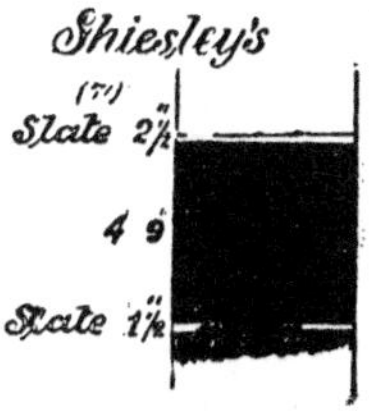

Shiesley's (71) Slate 2½" 4 9 Slate 1½"

Roof slate,	. . . . .	0′ 2″ to 0′ 3″
Coal,	. . . .	4 9
Black slate,	. . . . .	0 1 to 0 2
Bottom coal in floor.		

Of the four foot nine inch bench worked the upper three feet nine inches are not so hard nor so lustrous as the lower one foot.

A specimen of the upper part of this main bench yielded on analysis (M'Creath):

"Water,	1.600
Volatile matter,	30.700
Fixed carbon,	63.791
Sulphur,	.639
Ash,	3.270
	100.000

Coke, per cent., 67.70. Color of Ash, red.

The coal is resinous, compact, hard, slightly iridescent, with iron pyrites in veins."

A specimen of the lower part of this main bench of coal forwarded to Mr. M'Creath, yielded on analysis:

"Water,	1.480
Volatile matter,	29.220
Fixed carbon,	65.022
Sulphur,	.608
Ash,	3.670
	100.000

Coke, per cent., 69.30. Color of Ash, gray, with red tinge.

The coal has a resinous lustre, is very hard, and with much iron pyrites."

An old opening made many years ago on Fuller's Hill, one and a half miles east of Reynoldsville, is now fallen shut; but everything about the outcrop indicates the accuracy of the statement that the Lower Freeport bed showed its full thickness at that place.

At a new opening made by Carrier and Wilson, on their property, three miles east of Reynoldsville, on Soldier run, the coal shows (Fig. 72) as follows:

Fig. 72.

Carrier & Wilson.

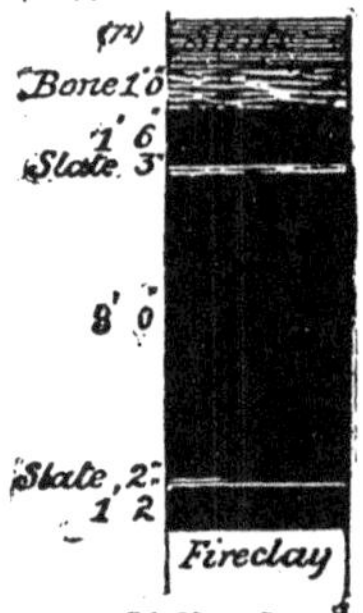

Roof, carbonated clay slates.		
Bony coal,	1′	
Coal, rotten,	1	6″
Slate,	0	3
Coal,	8	
Slate,	0	2
Coal,	1	2
Fire-clay floor.		

This opening, at the time of the examination, was only driven in a short distance from the outcrop; and in the partially decomposed mass it was difficult to judge just what amount of good coal the bed would yield. Appearances would indicate that the lower bench of fourteen inches, and the lower five feet of the eight foot bench will prove to be good coal.

At a natural opening on the same outcrop (De Lorme's) in a ravine some hundreds of yards to the eastward, four feet of the coal shows hard, bright, and apparently fairly free from iron pyrites.

No fair specimens could be obtained for analysis: but there is every indication that the coal will correspond closely to that of Seley's and Sprague's mines, opened in the same bed, and not far away.

The Lower Freeport Bed is opened at the Seley mine, three miles east of Reynoldsville, on Soldier Run. The coal measured in the mine, (Fig. 73,) as follows:—

Fig. 73.

Seley.

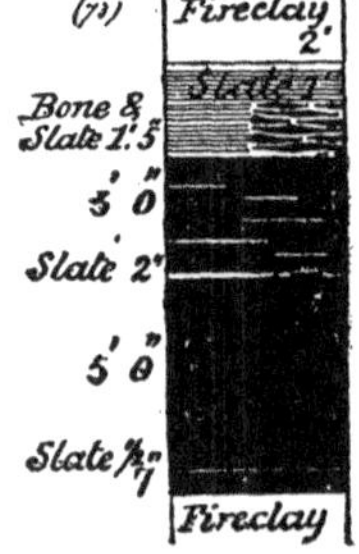

Fire-clay roof,	2′	0″
Rotten clay slates,	1	0
Bony coal and slate,	1	5
Coal, with slate layers,	3	0
Black slate. 0′ 1½″ to	0	3
Coal,	5	0
Black slate,	0	0½
Coal,	0	7
Fire-clay floor.		

Of this mass the workable coal is confined to the lower seven inch and five foot benches. Another measurement, in a different part of the mine, showed the lower seven inch coal bench increased to 14 inches in thickness, the small parting the same as before, and the main five foot bench of coal reduced to about four feet of good coal. Such variations as this, however, are found in almost every mine which has worked to any considerable extent this Lower Freeport coal bed in the Reynoldsville Gas coal basin.

A specimen of the upper part of this five foot bench of coal forwarded to Mr. M'Creath for analysis yielded:—

"Water,	0.850
Volatile matter,	31.200
Fixed carbon,	59.882
Sulphur,	1.368
Ash,	6.700
	100.000

Coke, per cent., 67.95. Color of Ash, dirty gray, with red tinge.

The coal has a dull, rusty appearance, is hard, iridescent, and contains pyrites in veins."

The lower part of this five foot bench of coal yielded, on analysis, (M'Creath):—

"Water,	1.040
Volatile matter,	31.610
Fixed carbon,	62.464
Sulphur,	.736
Ash,	4.150
	100.000

Coke, per cent., 67.35. Color of Ash, gray, with red tinge.

The coal has a dead lustre, is hard, iridescent, and contains small scales of sulphate of lime."

The lower bench, seven to fourteen inches in thickness, yielded on analysis (M'Creath):

"Water,	0.960
Volatile matter,	32.320
Fixed carbon,	58.640

Sulphur,	1.230
Ash,	6.850
	100.000

Coke, per cent., 66.72. Color of Ash, gray, with pink tinge. The coal has a dull lustre, is hard and compact, iridescent, and contains pyrites in quantity."

It may be mentioned that the coal from this lower bench is preferred by blacksmiths in their work.

An analysis of the ash of Seley's coal yielded (M'Creath):

"Silica,	1.860
Oxide of iron,	.220
Alumina,	1.760
Lime,	.060
Magnesia,	.162
Phosphoric acid,	——
Sulphur,	——
Percentage of ash in coal,	4.150

The Lower Freeport coal is opened at Sprague's mine, on Clayton run, a branch of Mix run, three miles east of Reynoldsville. The coal as measured in the mine shows (Fig. 74) as follows:

Fig. 74.

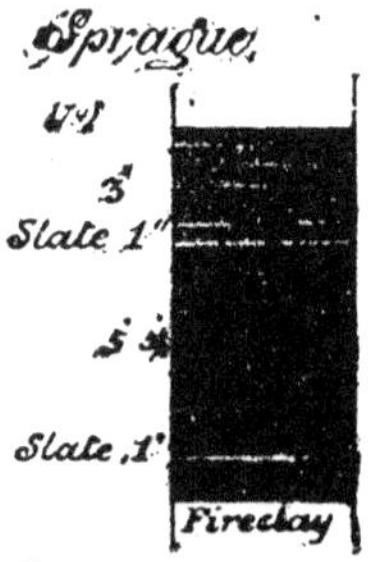

Sprague.
U.S
3'
Slate 1"
5'
Slate 1"
Fireclay

Coal in roof.			
Slate,			0' 1''
Coal,			5 5¼
Slate,	½'' to		0 2
Coal,	11 to		1 0
Fire-clay floor, soft.			

The upper part of the bed does not show in the mine, but to judge from the outcrop at the mine mouth, the same mass of coal, slate and clay exists on top of the main bench as is found at Seley's mine.

The lower bench of coal, the twelve inch bench, is preferred by blacksmiths for their fires.

A specimen of the main bench (five feet five and a half inches) of Sprague's coal, yielded on analysis (M'Creath):

"Water,	1.430
Volatile matter,	31.940
Fixed carbon,	62.109

Sulphur, - - - - - - -	.531
Ash, - - - - - - - -	3.990
	100.000

Coke per cent., 66.63. Color of Ash, cream.

The coal is shining on outside, resinous on fresh fracture with little iron pyrites."

These complete analyses of the Sprague and Seley Mi coals show that the Lower Freeport Coal Bed through this re gion retains its excellent character, and the sections show tha it has not fallen off in size.

With reference to the coals above and below the Lower Freeport Coal Bed in this region, though they are not very completely opened, yet sufficient has been done in picking open the outcrop to show roughly the general thickness.

The benches above the Lower Freeport Coal are nowhere opened, but show themselves on the hillsides.

On Seley's place, 95 feet below the Big Bed, a small coal has been picked into. It shows 17 inches of coal on the outcrop, and may run up, perhaps, to 30 inches of coal when under cover.

Thirty feet below this coal, and 125 feet below the Big Bed, a small coal is in the bed of Soldier's Run. This section corresponds with that at Reynoldsville; and the small size of these lower beds where opened at both places gives very little reason to hope that any of these beds will yield workable coal in the region between Reynoldsville and Seley's mine.

The Lower Freeport Coal is opened and worked at M'Creight's Mine, three miles South-east o Reynoldsville, on Mix Run. The coal a measured in the main entry of the min shows (Fig. 75) as follows:

Fig. 75.

Roof, black slate, - - - -	0′ 6″
Bony coal, - - - - - -	1 0
Coal, - - - - - - - -	1 9
Small slate parting.	
Coal, - - - - - - - -	3 0
Floor, soft fire-clay.	

Of this four foot nine inch bench of coal, the lower one foo is tender and friable, but good; the upper three feet nine inche of the bench being hard, bright and good coal.

Only one horse of fire-clay shows in the mine, cutting the coal down to three and three and one-half feet in thickness; the floor remaining smooth and regular. The mine is not worked for shipment to market, but only in a small way for local use.

The Strouse mine is opened on the Lower Freeport coal bed, three miles south south-east of Reynoldsville, on the waters of Trout run. The coal measured in the mine (Fig. 76) in one place, as follows:—

Fig. 76.

Strouse.

Slate & Bone (74)

5' 6"

Slate

Fire Clay

Roof, bony coal and slate.

Coal, ...	4′	0″
Small slate parting.		
Coal, ...	1	8

Fire-clay floor.

Where measured in another place the coal was four feet six inches in thickness, with a small fire-clay parting one foot above the floor. In still another place (all within 100 yards) the bed showed fire-clay roof, two feet: coal, with slate and bone coal, 28 inches: coal, five feet. Yet so irregular was the bed that in a short distance this 28 inch bench on top was entirely cut out, and the fire-clay roof rested directly upon the main bench of coal. This disturbed mine is quite in contrast to the usually regular mines working this Lower Freeport bed. The fire-clay horsebacks which are found nearly at the mouth of the main entry are also found almost persistently throughout its whole extent: the coal is pinched down in places to two feet of crushed and twisted coal, but is in no place entirely cut off.

The rolls are very numerous: both floor and roof being very uneasy and uneven.

A specimen of this coal forwarded to Mr. M'Creath for analysis yielded:—

"Water, - - - - - - -	1.300
Volatile matter, - - - - - -	30.220
Fixed carbon, - - - - - -	63.617
Sulphur, - - - - - - -	.763
Ash, - - - - - - - -	4.100
	100.000

Coke, per cent., 68.48. Color of Ash, with red tinge. gray,

The coal is bright, shining, clean, compact, with veins c pyrites and charcoal."

The bench of the Lower Freeport Bed is much less distinc at this disturbed Strouse mine: and for several miles soutl west of this point the hillsides fail to show the broad an clearly defined bench of the Big Coal sweeping along the sid of the valleys as it does in the vicinity of Reynoldsville a at other points where the bed is large and undisturbed. F some miles southwest there are but few openings: and the are in many cases on the Upper Freeport Bed. The Low Freeport Bed, however, resumes its full thickness and characte again when it reaches P. Hawk's and Hum's mines, to th north-west of Punxatawney.

At Phillippi's mine, three and a half miles south of Rey noldsville, the coal shows, where measured i the mine, (Fig. 77) as follows:

Fig. 77.

Phillippi

Roof, massive grayish sandstone.		
Soft clay slate and fire-clay,	. .	1′ 3½′
Coal, showing	. . . . . .	4
Bottom not seen.		

The mine runs in west 10° north for a shoı distance, and is partially filled with water, mak ing the dip to the north-west. It is said that another foot c good coal underlies that seen in the mine, making full five fe of good coal in all. A poor hematite ore, in reddish clay, i reported as underlying the coal, but was not seen. The coa though only seen near to the crop, looks good and hard, wit only a moderate show of iron pyrites, and no regular persister slate partings.

About 50 feet above Phillippi's mine, there is a small outcrc of black slate and coal smut; and also a small black slate sho on the road side, 30 feet below the mine. The measures overlyin the Phillippi mine are imperfectly shown, but are mainly gra ish sandy slates and thin sandstones. The coal is apparentl the Upper Freeport Coal Bed.

On Sheafer's place, four miles south of Reynoldsville, the is an old opening made on the outcrop of a coal, now fallen e tirely shut, which is reported to have shown five feet of goo coal when opened up. The roof is of massive sandstone, i ayers of from two to twelve inches in thickness, and the floor annot be seen. The hill rises 100 feet above this coal, the surace covered over with pieces of sandstone, fine grained. No ench shows on the hill side above the opening; but the openng itself is on a quite well defined bench. The sandstone ieces resemble closely those showing on the hill sides above e Lower Freeport coal openings at Reynoldsville; and this, th the appearance and size of the coal, and the level of the tcrop with reference to the other openings, indicates that his is the Lower Freeport Coal Bed outcrop.

In a ravine on Kroh's property, about one mile from Sheafer's, here is a natural opening, which shows, though somewhat imerfectly, the presence of a large coal bed, bright and hard at ts exposed outcrop, and with a micaceous sandstone roof resting directly upon it. This also is an outcrop of the Lower Freeport Coal Bed.

Secrist's mine is opened and worked about 4 miles south-west of Reynoldsville, on the waters of Trout Run. The coal, as measured in the mine, shows as follows, (Fig. 78):

Fig. 78.

Secrists.

(78) S.S.

4' 0"

Sandstone roof.	
Coal, - - - - - -	4'
Fire-clay floor.	

The massive gray sandstone rests directly on top of the coal. Where measured, only 15 feet in from he crop, the coal looked slaty, not in regularly formed and peristent slate layers, but in an apparent tendency of the whole ed to run into poor and ashy coal. The mine runs in southvest and drains, giving a local rise in that direction. The neasures above the coal are:—

Hill top.	
Shales and slate, - - - - -	25
Small black slate crop.	
Shales, - - - - - - - -	35
Thin sandstones, - - - - -	30
Coal, Secrists'.	

The bed, from appearance and level, is apparently the Upper Freeport Coal Bed.

After leaving the Woodward Reynolds coal mine, near Reynoldsville, and going south-west, the outcrop of the Lower

Freeport Coal Bed is seen plainly on the hillside at Dontilh's Mills, 160 feet above the stream; and the limestone has beer opened up under it as at the Woodward Reynold's mine. The limestone proved ferruginous and the quarry was abandoned. The hill rises 65 feet above the bench of the Big Bed, made up of loose shales and thin sandstones, but shows no marked bench This is the north-west outcrop of the Lower Freeport Bed i this part of the basin; Trout Run cutting out a broad valle and the sharp rise of the coal to the subordinate anticlinal ax (already described) carrying the coal over the tops of the hill north-west of Trout Run valley.

A small spring comes out of the hill just above water leve at Dontilh's Mills, marking the position of the small coal whicl at Reynoldsville is in the bed of the Sandy Lick creek.

The Hillis mine is opened about two miles south south-wes of Reynoldsville, on the waters of Trout run The coal as measured in the drift shows, (Fig 79,) as follows:—

Fig. 79.

Roof fire-clay.	
Coal, ...	4' 6"
Fire-clay floor.	

The opening when examined was only drivei in a short distance from the outcrop, and th roof was not yet regular: loose sandstone pieces coming dowı with the clayey matter resting on top of the coal. The hii rises 45 feet above the coal, and the measures, though not show ing clearly, seem from the surface stuff to be mainly made u of thin sandstones.

The Syphrit mine is opened not far from the Hillis mine, on th same bed. The coal where measured in the mine shows, (Fig 80,) as follows:—

Fig. 80.

Roof, sandstone.		
Yellowish clay and bone coal, .	0'	6"
Coal, ...	2	0
Slate, ...	0	0½
Coal, ...	1	1
Clay parting, ...	0	1
Coal, ...	0	7
Soft fire-clay floor.		

ThisTmeasurement was made in the main entry, 30 yards i from the outcrop, and under 30 to 40 feet of cover: but th coal was still soft and with much infiltrated clay on the face. At the extreme end of the heading it had improved in hardness: and seemed quite free from pyrites.

The Norris mine is opened three miles south-west of Reynoldsville. The coal where measured in the mine shows, (Fig. 81,) as follows:—

Fig. 81.

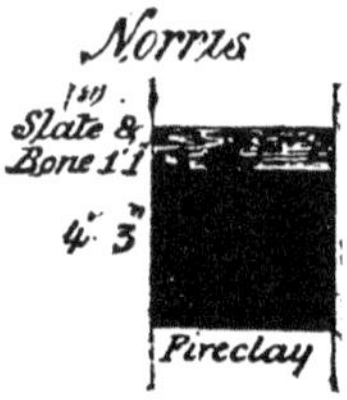

Norris

Roof, not seen.		
Slate and bone coal, . . .	1'	1''
Coal, ...	4	3
Soft fire-clay floor.		

The coal where measured was under sufficient cover but the roof was poor and leaky, so that the coal was soft, and stained with iron rust.

Some lumps of excellent carbonate iron ore which were lying at the mine mouth appeared to have come from just above the coal.

The hill rises 50 feet above and shows small sandstone pieces everywhere on the surface. The coal is rising rapidly to the North-west where opened.

At D. Brown's Mine, five miles south-west of Reynoldsville, coal is opened and worked. The coal as measured in the mine shows (Fig. 82) as follows:

Fig. 82.

D. Brown's

Roof, clay slate and fire-clay.	
Coal, ...	5' 0''
Floor, fire-clay.	

Though the mine is not driven in very far, the coal looks good. The hill only rises high enough above the mine to afford cover for the coal. The measures are here rising rapidly to the north-west, and this hill marks the outcrop edge of this coal bed on the north-west side of the basin here; for it shoots out to daylight finally on the west side of the hill.

One hundred and fifty-two feet below this D. Brown Coal Mine, on property of Mr. Jenks, a three-foot coal bed has been opened up, with a black slate roof; floor not seen. Mr. Fagen reports a sandstone, fine-grained and conglomeritic, as overlying the coal.

Mr. Whitesel has also opened up the coal upon his lands, near by.

About two miles south-west of this Mine, an imperfect vertical section of the measures above water level was made on Hog Shanty Run. The section gives (Fig. 83) as follows:

Fig. 83.

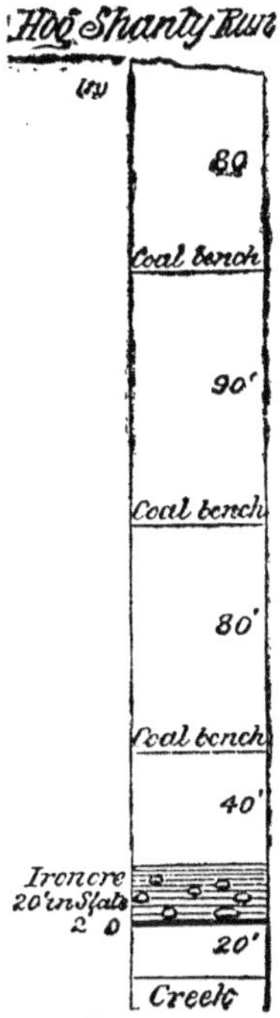

Hog Shanty Run

Hill top.*	
Concealed measures, - - - - -	80′
Bench.	
Concealed measures, - - - - -	90
Bench.	
Concealed measures, - - - - -	80
Bench.	
Concealed measures, - - - - -	40
Slates, with layers of iron ore balls, -	20
Coal.	
Concealed measures, - - - - -	20
Creek level.	

Of the coals and benches in the above 330 feet of measures, the lower coal is opened on the outcrop and shows two feet of good looking coal, resting on a hard fire-clay floor. Over the coal there are 20 feet of dark colored slates and shales. Much carbonate iron ore is found in this mass, mainly as nodular masses and partly as one and one-half to two inch plates. The shales were not completely cut down on the outcrop face, and no one regular and continuous workable bed of carbonate ore was seen; but the aggregate quantity of ore in these shales is clearly very considerable. It is said that a regular bed of carbonate iron ore, many feet in thickness, was opened up some years ago in this neighborhood; but it is now covered up and could not be found when the region was examined.

A specimen of this iron ore from Hog Shanty Run, forwarded to the laboratory of the Survey, in Harrisburg, yielded on analysis (M'Creath):

"Iron, - - - - - - - -	34.000
Sulphur, - - - - - - -	.355
Phosphorus, - - - - - -	.202
Insoluble residue, - - - - -	11.390

The ore is a carbonate, exceedingly hard and compact, minutely crystalline, of a dark gray color, with conchoidal fracture; and showing crystals of pyrites."

The bench 80 feet above the stream had not been opened. The bench 160 feet above the stream had only been just picked into at the time that the examination was made—enough to show a fire-clay floor to the coal, but without having yet developed the true roof, only loose matter resting on the coal. This bench is quite marked; fallen trees occasionally bringing up pieces of coal along it. The upper bench, 250 feet above the stream, had not been opened. Nothing shows from that to the hill top.

Wachob's mine is opened about 5 miles north north-west of Punxatawney. The coal as measured in the mine shows (Fig. 84) as follows:

Fig. 84.

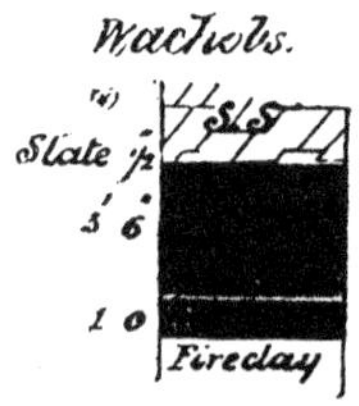

Roof, sandstone.			
Black slate, - - -	0	0½″ to 1′	
Coal, - - - - - -	3	6	
Slate, persistent, - -	0	0½	to 1
Coal, - - - - - -	1	0	
Soft fire-clay floor.			

The coal is rising to the north-west. The mine is only driven in for about 40 yards, and has been worked on a small scale merely to supply the local demand. Gray slates and thin sandstones overlie the coal for 100 feet, and at 70 feet above the coal there is a small black slate outcrop

The coal mines out well, firm and in blocks.

A specimen forwarded to Mr. M'Creath for analysis, yielded:

"Water, - - - - - - -	1.300
Volatile matter, - - - - -	32.570
Fixed carbon, - - - - -	62.567
Sulphur, - - - - - - -	1.023
Ash, - - - - - - - -	2.540
	100.000

Coke, per cent., 66.13. Color of Ash, red.

The coal has a dead lustre on outside; on fresh fracture, a bright lustre, and is strongly iridescent."

At Elbles', five miles north of Punxatawney, two coal beds have been opened, 50 feet apart. They were not worked at the time of the examination and could not be accurately measured.

The upper bed shows a black slate roof, a fire-clay floor, and apparently a four foot coal. Masses of sandstone, fine grained and conglomeritic, lie around the mine mouth, and above the coal.

Fifty feet below this mine is the old drift on the lower bed. The drift runs in east 20° south and is filled with water: indicating a south-east dip to the measures here. In the main entry, near the mine mouth, the show of coal is:

Heavy smooth black slate roof.	
Coal, ...	4′
Fire-clay floor.	

The coal is reported as used and desired by blacksmiths, and as bearing an excellent name for domestic use. This opening lies 20 feet above the level of the small stream in the ravine.

About a half mile south-west of Elble's, Zeitler's mine is opened up, apparently on the lower bed of the two showing at Elble's. The coal, as measured in the mine, shows (Fig. 85) as follows:

Fig. 85.

ZEITLER.

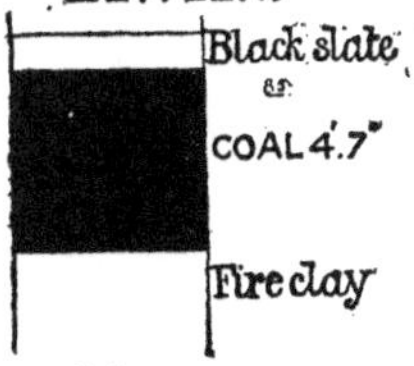

Black slate

COAL 4′.7″

Fire clay

Black slate roof, smooth, . .	1′	0′
Coal, good, ...	4	7
Fire-clay floor.		

For a considerable distance in the mine the coal is covered over with a coating of yellowish clay. As the black slate roof is tough and good, and the cover ample, this infiltration probably comes from an horse of fire-clay, located somewhere up the rise of the coal. Free from this infiltrated matter, the coal is good, bright and clean; the upper six inches being duller in lustre and harder.

The mine runs in west and drains; showing a rise to the north-west, a south-east dip.

Seventy-two feet higher on the hill, but not directly above, is Zeitler's opening on the upper bed. It is now not worked and has fallen shut; but the main entry has six foot timbers in it, indicating probably a good sized bed. Above this bed there are gray thin bedded slates and shales for 150 feet, with no outcrops or benches showing.

At the John Henry mine not far north of Zeitler's, the coal is opened up and shows where measured in the mine, (Fig. 86) as follows:—

Fig. 86.

J. Henry.

Slate

4′ 6″

Fireclay

Roof black and grey slate.	
Coal, ...	4′ 6″
Fire-clay floor?	

The coal is good, with no bone or persistent slate showing. The bottom nine inches are very hard, then two feet nine inches of moderately firm coal, then one foot of soft friable coal on top. The coal is dipping very gently to the south-east, is mined to the dip, and drained by a trench.

Scheldt's mine is opened just north of Henry's. Nothing shows.

At Wolfe's mine the coal shows:

Black slate roof.	
Coal, ...	4′ 6″
Fire-clay floor.	

The coal shows the same sub-divisions in character as mentioned as showing in the J. Henry mine, and rises very slightly to the north-west. Grayish slates make up the overlying measures.

At Smause's mine, two and a-half miles north north-east of Punxatawney the coal shows:

Sandstone.	
Black slate roof, . . .	3′ 2″
Coal, (the upper six inches with a cannel structure,)	3 6
Fire-clay floor.	

The coal is rising gently to the north-west. The coal looks hard and good. The measures overlying, to the hill top (70 feet) are made up mainly of sandstones.

The Lower Freeport Coal Bed is opened and worked at the Powell Hawk mine, two and a-half miles north of Punxatawney. The coal as measured in the mine shows, (Fig. 87) as follows:—

Fig. 87.
Hawk.

Roof, hard black slate.	
Coal, with some slaty layers and bony coal, ...	1′ 0″ to 2′ 0″
Slate persistent, ...	0½ to 0 2
Coal, ...	4 6 to 5 0
Fire-clay floor.	

The lower five foot bench of this coal yields a very handsome, hard, clean, bright coal. The upper two foot bench, with its slate layers and bone coal is mined for local use, but could never be shipped to market.

A specimen of the main bench yielded, on analysis, (M'Creath):

"Water, - - - - - - -	0.950
Volatile matter, - - - - - -	33.550
Fixed carbon, - - - - - -	60.523
Sulphur, - - - - - - -	1.167
Ash, - - - - - - - - -	3.810
	100.000

Coke, per cent., 65.50. Color of Ash, reddish brown.

The coal has a shining lustre, is hard and compact, with mineral charcoal and much iron pyrites."

Gray slates and thin sandstones overlie the coal, and a small bench shows on the hill side some 50 feet above it.

The Lower Freeport Coal Bed has here regained its character and full size, as displayed by it in the mines at Reynoldsville, and eastward of that point. The sections and descriptions of the different mines have shown no single opening between M'-Creight's mine and Powell Hawk's mine, where the coal rises to this full thickness of good coal. The extraordinary size and excellence of this coal at Reynoldsville and north-west of Punxatawney, must be put down as an abnormal and fortunate condition of the bed; not to be relied upon or expected in other parts of the basin.

At the Campbell mine, one-half mile south of the Hawk mine, the same coal is opened and worked. The coal shows, where measured in the mine, (Fig. 88) as follows:

Fig. 88.

Sandstone, massive.	
Roof, fire-clay,	2' to 3' 0'
Soft slate,	0 6
Bone coal,	0 7
Coal,	4 8
Floor, not seen.	

The dip appeared to be to the north-west, and the mine was filled with water. The coal was measured near the outcrop, where it was still partially disintegrated and could not afford a perfect measurement.

About two-thirds of a mile south of Campbell's mine, W. Mitchell has opened up the outcrop of a coal only sufficiently to show one foot of coal, with blue slates overlying, and then blue carbonate of iron above.

The Lower Freeport Coal Bed is opened and worked at the J. Thomas mine, two and a half miles north of Punxatawney. The coal, as measured in the mine, shows (Fig. 89) as follows:

Fig. 89.

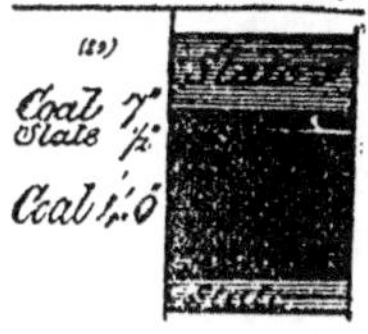

Roof, black slate,	3' 0"
Coal,	0 7
Slate,	0 0½
Coal,	4 0
Black slate and coal in bottom.	

At other points in the same mine, the bed measures six feet from roof to floor, the floor being fire-clay. The coal is rising very gently to the north-west. Gray slates five feet thick overlie the black slates of the roof, and then sandstone.

The coal from the main bench is bright and shining, but rather friable.

A specimen of the coal forwarded for analysis, yielded (M'-Creath):

"Water,	0.950
Volatile matter,	31.590
Fixed carbon,	60.520
Sulphur,	1.440
Ash,	5.500
	100.000

Coke, per cent., 67.46. Color of Ash, dirty gray.

The coal is bright, shining, compact, with many small veins of charcoal and pyrites."

The Upper Bed, 75 feet above the Lower Freeport Coal Bed, is opened at the Weaver mine, 3 miles north north-west of Punxatawney. The coal, as measured in the mine, shows (Fig. 90) as follows:

Fig. 90. Weaver

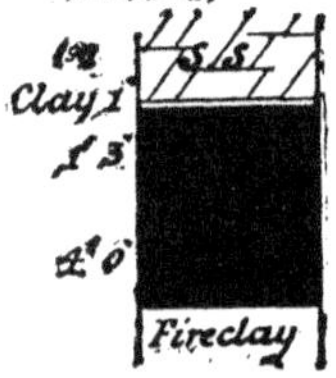

Roof, massive sandstone.		
Rotten clay slate,	0'	1"
Coal, poor and rotten,	1	3
Coal, good,	4	0
Fire-clay floor.		

The coal is rising to the north-west gently. As usual the massive sandstone roof is cracked in places, and the face of the coal is much stained with infiltrated clay; but the coal itself is hard and firm.

The mine in roof, floor, size and character resembles the openings on the Upper Freeport Bed, 6 miles to the north-east (Phillippi, &c.) already described; speaking very well for the regularity of this bed through all the intervening country.

A specimen of the coal forwarded for analysis yielded (M'-Creath):

"Water,	1.000
Volatile matter,	33.200
Fixed carbon,	59.428

Sulpnur, - - - - - -	2.042
Ash, - - - - - - - - -	4.330
	100,000

Coke, per cent., 65.80. Color of Ash, brown, with reddish tint.

The coal has a dull lustre, is coated with silt, friable, iridescent, with charcoal and pyrites."

About 80 feet above the bed there is a bench on the hill side: and just above the bench massive sandstone boulders on the surface indicate the position of the Mahoning sandstone.

The same coal bed has been opened and worked at Conrad's Mine, one-half mile east of Weaver's. The coal as measured in the mine shows (Fig. 91) as follows:

Fig. 91.

Roof, sandstone.		
Yellowish clay, - - - - - -	0'	1"
Coal, soft, bony in places, - - -	1	0
Coal, - - - - - - - - - -	4	6
Fire-clay floor.		

There is only one small slate parting in the lower four foot six inch bench of coal, and the coal looks hard, good and clean. It shows considerable pyrites. The mine is opened just above the level of the small run and the coal is rising gently to the north-west. Thin bedded sandstones and silicious slates overlie the sandstone roof.

The opening at "Smith's Old Bank" is now fallen shut and could not be measured. The same massive sandstone roof shows in place. It lies not far from Conrad's mine.

North of these openings about two miles, or about five miles north north-west of Punxsutawney, there is a natural opening of the coal in a ravine on Kessler's place. The coal, which is but imperfectly exposed, shows (Fig. 92) as follows:

Fig. 92.

Dark clay slates.		
Roof, black slate, - - - - -	2'	0"
Coal, - - - - - - - - - -	3	6
Black slate, persistent, - - ½" to	0	1
Coal, - - - - - - - - - -	1	3
Coal in bottom.		

The coal is hard and bright and seems to be

of excellent character. The roof slate is for the lower six inches a cannel slate; above that a black slate.

A small coal has been opened 40 feet above this Kessler opening. It is now fallen shut, and is reported a "3 foot bed." Sandstone pieces cover the hill side between these beds, and above the upper one for 50 feet to the hill top. The Big Bed is 150 feet above the stream in the valley below, and the upper bed 190 feet. These beds are, apparently, the Lower Freeport coal, and the coal (Middle Freeport) 40 feet above. This middle coal is evidently very irregular, in places existing as a workable bed and again running down to almost nothing. It has already been mentioned as making a well defined bench on the hill side above Seley's bank, three and a half miles east of Reynoldsville, 35 to 40 feet above the Lower Freeport bed.

Fig. 93.

Elk Run

Slates 60'
S.S. 10'
Slates 55'
Coal
S.S. 40'
Coal
Slates
S.S. 155'
Creek

On this Kessler hill no marked bench shows between the Big Bed and the stream, 150 feet.

An imperfect vertical section on the west side of Elk creek, opposite Kessler's, shows (Fig. 93) as follows:

Hill top.	
Slates,	60′
Sandstone, (Mahoning?)	10
Slates, grayish and ferruginous,	55
Bench of coal, Middle Freeport,	3
Sandstone, thin,	40
Bench of coal, Kessler, Lower Freeport.	
Gray shales and thin slates and sandstones, with no marked bench,	155
Level of Elk run.	

Pantall's mine is opened and worked four and a half miles north-west of Punxatawney. The coal as measured in the mine shows (Fig. 94) as follows:

Fig. 94.

Pantalls

Slate 4'
1'8"
Clay
3'0
1'0
Fireclay

Roof, black slate, hard,	4′	0″
Coal,	1	8
Clay parting, not persistent,	0	2
Coal,	3	0
Slate parting.	0	0½
Coal,	1	0
Fire-clay floor.		

The coal is rising to the north-west. The mine is only worked to supply a small local demand and is not driven in far; so far as worked the coal looks very well, but the horses of fire-clay are numerous and troublesome. This mine, like the Wachob and D. Brown mines, is over on the north-west side of the basin, and the coal has risen high in coming up to the anticlinal axis which lies west of the mine. In fact, this Pantall coal is already well into the hill tops and the Straighthoof outcrop, about one mile north-west of Pantall's, probably marks about the western edge of the basin, at this point, for the Freeport coals.

A specimen of coal from Pantall's mine yielded on analysis (M'Creath):

"Water,	1.100
Volatile matter,	31.170
Fixed carbon,	63.544
Sulphur,	1.016
Ash,	3.170
	100.000

Coke, per cent., 67.73. Color of Ash, yellowish brown.

The coal has a dull lustre, is hard, coated with silt, shows charcoal and pyrites, and is slightly iridescent."

M'Kee's mine is opened and worked three and a-half miles north-west of Punxatawney. The coal as measured in the mine shows (Fig. 95) as follows:

Fig. 95.

Roof, black slate.	
Bone coal and slate,	1′ 0″
Coal,	5 1
Fire-clay, soft, in floor.	

The coal rises to the north-west. Although the main entry is driven in for over 50 yards the coal is somewhat rotten and stained with infiltrated clay.

A specimen of the coal forwarded for analysis, yielded (M'Creath):

"Water,	1.050
Volatile matter,	33.150
Fixed carbon,	58.405

Sulphur,	1.295
Ash,	6.100
	100.000

Coke, per cent., 65.80. Color of Ash, gray.

The coal is shining, compact, somewhat slaty, with considerable iron pyrites."

The hill rises 62 feet above the M'Kee mine and shows a small black slate crop on top, but without any coal smut.

At the J.Smith mine, half a mile from M'Kee's, the coal shows:

Roof, black slate.	
Coal,	4′ 6″
Fire-clay floor.	

The coal does not show hard or firm where examined, and is rising to the north-west.

The Wingert mine is opened and worked three miles north-west of Punxatawney. The coal as measured in the mine shows (Fig. 96) as follows:

Fig. 96.

Roof, black slate.		
Bony coal and slate,	2′	0″
Coal,	3	3
Slate,	0	0
Coal,	2	0
Fire-clay floor soft.		

The coal looks good and bright, the upper bench, above the small slate parting, being hard and mining out well in lumps: while the lower bench is columnar in structure and friable, but good. The mine is almost entirely free from the clay "horses" which were noted in the Pantall and M'Kee mines. The hill rises 75 feet above the mine.

A specimen of this Wingert coal yielded, on analysis, (M'-Creath):

"Water,	1.150
Volatile matter,	32.070
Fixed carbon,	60.428
Sulphur,	1.702
Ash,	4.650
	100.000

Coke, per cent., 66.78. Color of Ash, fawn.

The coal is bright, friable, with pyrites and Oxide of Iron."

At Dr. Kurtz's mine, the coal shows:

Roof, black slate.	
Coal, - - - - - - -	5′ 7′
Soft fire-clay floor.	

The coal is hard, bright, clean and quite free from pyrites. Over the coal there are 15 feet of ferruginous slates and thin sandstones.

At the J. B. Morris Mine (No. 1) the coal shows:

Grayish blue sandy slates.	
Coal, - - - - - - -	5′ 3″
Fire-clay floor,	

The Jones Mine is opened and worked near the Morris Mine and two miles north-west of Punxatawney. The coal shows where measured in the mine as follows (Fig. 97):

Fig. 97.

Jones.

Roof, sandy slate.	
Coal, - - - - - - - -	5′ 0′
Fire-clay floor.	

In the black slate roof occur some layers of black silicious iron ore.

Fifty-six feet above the coal is a bench, apparently of coal. A bastard limestone and ore are reported as struck on this bench, and the well of the house is limestone water. This would be the Freeport limestone in place, with the Middle Freeport Coal resting on top of it.

At Carmalt's Mine the coal is opened at water level of the creek, and the bed shows, as imperfectly seen:

Roof, black slate, - - - -	2′ 0″
Slate and bone coal, - -	1 0
Coal, showing, - - - -	3 0
Bottom not seen.	

It is reported full four and a half feet of coal. The coal showing is good, bright and clean, and lies nearly flat. A slope was put down west of the stream and the coal worked from it; but it is now abandoned. Large blocks of massive sandstone show in the stream valley, washed down from above. At the Carmalt mine the hill rises 50 feet above the coal; but to the east of the mine the hill rises 155 feet above the mine. The measures are very little exposed, and seem to be mainly gray slates with sandstone layers, one small black slate show, without coal smut, appearing 130 feet above the mine.

Hum's mine is opened and worked about two-thirds of a mile north of Carmalt's and two and a half miles north-west of Punxatawney. The coal, as measured in the mine, shows (Fig. 98) as follows:

Fig. 98.

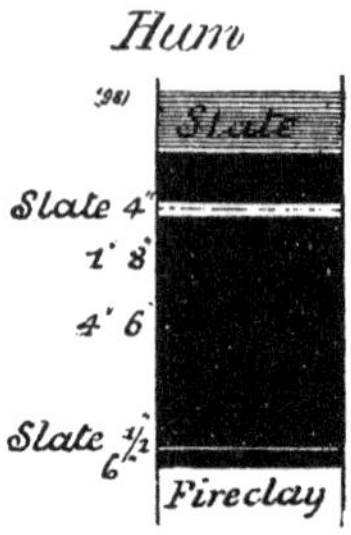

Hum

Coal in roof.		
Slate, ...	0'	4'
Coal, hard, ...	1	8
Coal, softer, ...	4	6
Slate parting, ...	0	0½
Coal, ...	0	6
Fire-clay floor, hard.		

Where measured as above the roof was not shown, and the roof coal was reported as "3 feet." Where measured in another part of the mine the roof was found to be a hard black slate; but in this case the mine was not cut down from roof to floor. When examined for the First Geological Survey of Pennsylvania, this bed, as showing here, was called an eight foot coal bed, and it may average for some distance that very great thickness. The coal is good, clean and handsome, and mines out well. The dip, as given by the workings, seems to be gentle to the south 10° west. This is one of the handsomest exhibitions of clean, bright coal to be found in the Reynoldsville Gas Coal Basin.

A specimen of the upper part of the main bench of Hum's coal, forwarded to Mr. M'Creath for analysis, yielded:

"Water, - - - - - - -	0.920
Volatile matter, - - - - -	35.440
Fixed carbon, - - - - -	59.962
Sulphur, - - - - - -	.848
Ash, - - - - - - - -	2.830
	100.000

Coke, per cent., 63.64. Color of Ash, brown, with reddish tinge.

The coal is bright, with glossy lustre, compact, with small amount of iron pyrites."

A specimen from the lower part of the main six foot bench yielded on analysis (M'Creath):

"Water, - - - - - -	1.000
Volatile matter, - - - - -	33.260

Fixed carbon,	63.081
Sulphur,	1.139
Ash,	1.520
	100.000

Coke, per cent., 65.74. Color of Ash, reddish brown.

The coal has a shining lustre, is somewhat compact, with a small amount of pyrites."

These analyses show that this coal is superior in quality as well as unusually great in size.

The J. B. Morris No. 2 mine is opened and worked about one-fourth of a mile from Hum's. The coal, as measured in the mine, shows (Fig. 99) as follows:

Fig. 99.
Morris 2

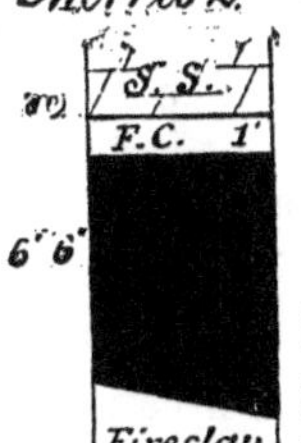

Gray sandstone.	
Roof, hard fire-clay,	1' 0"
Coal,	6' to 7
Fire-clay floor, soft.	

The coal is without any persistent slate partings, and is very hard and bright. The swelling in the size of the Lower Freeport Coal Bed embraces the region covered by these two openings; but it is, of course, local to this spot, and is not found in any other of the numerous mines already described.

Ruth's Mine is opened and worked about five miles west north-west of Punxatawney. The coal as measured in the mine shows (Fig. 100) as follows:

Fig. 100.
Ruth's

Slate
5' 3"
Fireclay

Roof, black slate.	
Coal,	5' 3"
Fire-clay floor.	

The coal is dipping to the south-east; and as it has very little cover over it where worked, shows weathered, dirty and with infiltrated clay on the faces of the coal.

A specimen forwarded for analysis yielded (M'Creath):

"Water,	1.060
Volatile matter,	34.140
Fixed carbon,	61.172
Sulphur,	.678
Ash	2.950
	100.000

Coke, per cent., 64.80. Color of Ash, cream.

The coal has a dead lustre, is compact, iridescent, hard, coated with silt, and shows charcoal and iron pyrites."

The Means Mine is opened about half a mile east of Ruth's. This opening is much troubled and pinched in the short extent of the workings, showing at one place fire-clay in the roof, at another place sandstone, and at a third black slate.

The coal only showed three feet six inches in all, of which the upper six inches were bone coal.

Another opening made along the same bench, some few hundred yards away, found the coal regular and undisturbed. The coal as measured in this new drift shows (Fig. 101) as follows:

Fig. 101.

Roof, black slate, partially decomposed.

Coal, - - - - - - - - -	4′	2′
Slate, - - - - - - - - -	0	0½
Coal, - - - - - - - - -	1	6

Floor not seen.

The coal looks hard, bright and good, and is rising in the mine to the north-west. The hill rises 75 feet above the mine, the measures made up apparently of soft, thin shales.

Anthony's Mine is opened and worked not far from Means, and about four and-one-half miles west north-west from Punxatawney, near the village of Pottersville. The coal as measured in the mine shows (Fig. 102) as follows:

Fig. 102.

Roof, black slate.

Black slate and bone coal, - - -	1′	0′
Coal, hard and good, - - - -	3	9
Slate, - - - - - - -	½″ to 0	1
Coal, - - - - - - - - -	0	6

Fire-clay floor,

In another place in the mine the three foot nine inch bench measured four feet of excellent coal.

A specimen of this Anthony Mine coal yielded on analysis (M'Creath):

"Water, - - - - - - -	0.950
Volatile matter, - - - - -	35.870
Fixed carbon, - - - - -	58.218
Sulphur, - - - - - -	2.302
Ash, - - - - - - - -	2.660
	100.000

Coke, per cent., 63.18. Color of Ash, red.

The coal is shining, somewhat friable, with many small veins of charcoal and pyrites."

A row of small springs coming out of the hillside 30 feet below the Ruth mine indicates the presence there of a coal or limestone; but no opening has been made.

About 75 feet above Anthony's mine, but not on the same hillside, a limestone has been opened up, three to four feet thick. When burned it yielded only a tolerable lime, the limestone carrying too much iron. No coal smut shows above it or below it. It is probably the Freeport limestone.

The Fourth Coal Basin continues on south-west from these Ruth, Pantall, Anthony, &c., mines, and coal is opened on many farms. But the mines are never reported as more than "4 feet" or "4½ feet" coals; and the abnormally great size and excellent gas coal character of these Freeport Coal Beds, as described above, is not apparently repeated to the south-west, in the district between the Ruth mine and the point, 10 to 12 miles to the south-west, where the steady sinking of the measures to the southward has carried the Freeport coals below water level, and has made the "barren measures," the country rock above water level.* These barren measures, made up mainly of a wilderness of greenish, olive colored and reddish shales, holding only small and worthless coals, continue as the surface rock south-west through Indiana county, until the Pittsburg Coal Bed, coming in on top of them, is found occupying the high lands north of Salzburg.

Oil boring at Punxatawney.

The following is the record of an oil boring made in 1864 on the Mahoning creek at Punxatawney. It is kindly furnished by Mr. P. W. Jenks, who reports it as "carelessly recorded:"

Soil,	10′	0″
Reddish sandstone,	5	0
Very hard sand rock, with iron ore mixed,	5	0
Black slate,	5	0
Slate,	6	6
Sand rock,	12	0
Slate rock,	18	0
Soft sand rock,	21	0

(*These barren measures, lying above the Mahoning sandstone, are at this point about 500 to 600 feet thick.)

Not recorded,	35′	6″
Show for oil, (118 feet below surface.)		
Not recorded,	8	0
Coal bed,	6	0
Slate,	22	0
Coal,	4	0
Slate rock,	2	0
Sandstone, glass sand,	1	0
Coal, called "Cannel coal,"	2	0
Not recorded,	5	0
Show for oil at 166 feet.		

Iron Ore and Limestone at Punxatawney.

At Clayville, one mile west of Punxatawney, Mr. Young examined the ore and limestone outcrops as imperfectly exposed on the bank of the Mahoning creek. The Freeport (?) limestone occupies the bed of the creek; then sandstone, one foot; fire-clay and ore balls, 6 feet; limestone and some ore, in all twelve feet. As is not unusual, the Freeport is carrying an upper layer of very ferruginous limestone, at times an iron ore. The ore splits easily and is of a greenish tint.

A specimen of this Freeport iron at Claysville, forwarded to Mr. M'Creath for analysis, yielded:

"Iron,	21.100
Sulphur,	.127
Phosphorus,	.493
Insoluble residue,	30.010

Though not rich in metallic iron, yet the low percentage of sulphur and phosphorus in the above ore would in all probability make it pay to work. When roasted it would yield an ore of sufficient richness in iron and of decidedly good quality.

It was not exposed sufficiently to get a safe average of its thickness, and it would well repay opening up to settle this point.

About one-quarter of a mile below the bridge the ore again shows in the slates, but more sandy than before. At the head of the bend of the creek (Sliding Rock bend) the ore again shows; but very sandy and poor. The fossiliferous calcareous shale and slate overlying the limestone and ore is very persistent, being found again down the Mahoning creek at Smicksburg. On top of this comes in sandstone, with thin slates overlying; and still higher in the hill is the Mahoning sandstone, the massive boulders of which fill the stream bed. On top of this come in the slates and shales of the barren measures mentioned above.

Coals on the Mahoning.

Going down the Mahoning creek for five miles (see Rogers' Final Report) below Punxatawney, the Lower Freeport coal is opened six feet thick, with six feet of shales covering it. At this place the Freeport limestone is in two layers, the uppermost twenty inches thick, of good blue carbonate; then two and a half inches of bright yellow clay; and underneath it twelve inches of limestone are exposed, which may be thicker. Slabs of gray carbonate of iron, one and a half inches thick, are found in the same quarry; and also a shaly, silicious, hard and heavy ferruginous semi-limestone stratum, recognizable also in the Smicksburg section. At Templeton, where the Lower Freeport coal (?) is only opened as a three and a half foot coal bed, with shale roof and slate floor, at the level of the creek, the Freeport limestone, 10 to 12 feet above it, is a band two and a half feet thick, floored by shelly sandstone, looking like a decomposed ferruginous rock. At another opening it appears in thin layers, thus:

Pure blue limestone, . .	0'	4'
Black slate, ...	0	3
Yellowish limestone, . .	0	4
Clay, ...	0	3
Limestone, ...	2	0
Slate.		

At Kinter's Mill, where the Fourth anticlinal axis crosses the Mahoning valley, the Freeport limestone just reaches water level. Above it is a small coal bed, the Upper Freeport coal, and high up towards the barren measures, which occupy all the high lands, are calcareous shales with a six inch bed of coal upon them. These barren measures occupy the whole basin south of Punxatawney, yielding no workable coal, and no large bed of limestone, but displaying the conglomerate Mahoning sandstone everywhere as their base rock. (Rogers' Final Report.)

Opposite Punxatawney, on the south side of the Mahoning creek, a small coal show is seen 130 feet above the creek; a bog ore deposit at 200 feet; and a small black slate outcrop, in the shales, at 280 feet above the Mahoning creek.

The coal openings in the Third Bituminous Coal Basin, east and south-east of Punxatawney, were examined to see if any of the exposures would exhibit a coal bed which could compare with the size and character of the handsome coal already discribed north-west of that place.

At Brown's mine, near the mouth of Stump creek, four miles east of Punxatawney, the coal shows:—

Roof, black slate, crumbly.	
Coal, - - - - - - - -	3′ 2″
Fire-clay floor.	

The coal seems friable; and is apparently dipping to the north-west.

At Andrew Bower's mine, one and a half miles south of Brown's, the coal shows:—

Heavy black slate roof.	
Coal, - - - - - - -	3′ to 3′ 6″
Fire-clay floor.	

Six inches above the floor is a persistent small black slate parting; and irregular small slates come in through the mass of the coal.

About four miles south-east from Punxatawney, in Indiana county, the G. Schlimmer mine is opened. The coal shows:—

Roof, black slate.	
Coal, - - - - - - - -	4′
Fire-clay floor.	

The coal shows slate partings, not persistent. Fifty feet of sandy slates underlie this mine, down to the level of Schlimmer's run, and through this whole mass are scattered nodular masses of carbonate iron ore in lumps of all sizes; but in no one place were the ore balls closely enough packed, or the ore plates thick enough to make a workable bed of iron ore. It is reported that a persistent bed of carbonate iron ore of considerable size lies directly under Schlimmer's coal bed; but it could not be seen when the mine was examined.

A specimen of this Schlimmer iron ore yielded, on analysis, (M'Creath):

"Iron, - - - - - - - -	26.500
Sulphur, - - - - - -	.141
Phosphorus, - - - - - - -	.149
Insoluble residue, - - - - - -	34.460

The ore is a carbonate, is hard, compact, silicious, of a bluish gray color, with conchoidal fracture."

Above the coal are thin gray slates, shale and sandstone. The same coal bed apparently is opened by Wolf, Smith, George Brooks, Spencer and M'Cullough.

At the Joel A. Ginter mine the coal shows:

Black slate roof.	
Coal, - - - - - - -	5′ 0′
Fire-clay floor, hard.	

The coal shows a few small black slate partings, not persistent, except one small half inch slate, six inches above the floor. The roof slate is of a peculiar smooth cannel slate character, very fossiliferous. The measures rise to the south-east.

The Geo. Graff mine is opened six miles south-east of Punxatawney, in Indiana county. The coal as measured in the mine shows, as follows:

Roof, fire-clay.		
Bone coal, ...	0′	6″
Coal, ...	1	7
Fire-clay, ...	0	0½
Coal, ...	3	0
Fire-clay floor.		

The mine runs in South 30° east and does not drain.

A specimen of the coal yielded, on analysis, (M'Creath).

"Water, - - - - - - -	1.050
Volatile matter, - - - - - -	29.730
Fixed carbon, - - - - - -	59.781
Sulphur, - - - - - - -	1.389
Ash, - - - - - - - -	8.050
	100.000

Coke, per cent., 69.22. Color of Ash, reddish brown.

The coal is of a dirty appearance, somewhat hard, with considerable iron pyrites."

Going south-east for four miles the measures are found rising up towards the Second Anticlinal axis; and at Andrew Pierce's place, 500 feet above the Mahoning creek, the hill crest is covered with massive sandstone boulders which show the Mahoning sandstone, here risen high up and shooting out into the air to arch over the Second Anticlinal axis.

A number of coals are opened in the vicinity of this place, all 100 to 200 feet below the Mahoning sandstone.

The M'Farland coal mine is opened in Canoe township, Indiana county, and shows (Fig. 103) as follows:

Fig. 103.

Roof, black slate.
Coal, - - - - - - 3′ 8′
Fire-clay floor.

There is not much cover to the coal where opened, and considerable infiltrated clay is found on the faces of the coal.

The coal is rising to the south-east.

A specimen of this coal forwarded to Mr. M'Creath for analysis yielded:

"Water, - - - - -	1.020
Volatile matter, - - - - -	30.190
Fixed carbon, - - - - -	57.943
Sulphur, - - - - - -	2.757
Ash, - - - - - - -	8.090
	100.000

Coke, per cent., 68.79. Color of Ash, cream.

The coal is bright, iridescent, columnar, with considerable iron pyrites."

The hill rises slowly 50 feet above the M'Farland mine, and at the top the surface of the ground, for some acres, is covered with small pieces of lean, sandy hematite iron ore. This same ore is also found about one-half of a mile north-east, on S. Pierce's place.

At the M'Cullough mine the coal shows:

Roof, black slate.
Coal, ... 4′ 0′
Fire-clay floor.

The coal, even near the outcrop, looks good and bright, but evidently carries considerable pyrites. Fifteen feet below the coal a carbonate iron ore was formerly opened up on the outcrop; but it is now fallen shut and cannot be measured.

On S. Pierce's place, near his house, is an old opening on a coal bed, now fallen shut. A shaft put down just alongside of the coal mine found limestone and carbonate ore, not more than 10 to 15 feet below the coal. No measurements could be made. This is on the north side of Wilson's run, near its head waters.

On the south side of the run, openings have been made on the outcrop, and the same measures were developed; the limestone being in its place under the coal, as before. The rocks both above and below the limestone are mainly gray slates. This is probably the Freeport limestone, carrying its usual variable ore on top.

A ferruginous clay slate, deep red in color, eight to ten feet thick, and in places even more, has been opened on White's place, in Indiana county, five miles south-east of Punxatawney, for an iron ore. An analysis shows it to contain not more than 10 per cent. of metallic iron; and it is, of course, entirely worthless. Above it lies a coal which has been opened and worked on the adjoining property, (North's) and is reported to have shown as a four foot bed of coal.

The analyses of these coals in the north-eastern end of Indiana county, show considerable sulphur, and also a decidedly large percentage of ash; indicating that the coals of the Third Bituminous Coal Basin, as previously found in the same basin farther to the north-east, are clearly inferior in quality to the coals from the Freeport Coal Beds, in the Fourth Coal Basin south-west of Reynoldsville. And the vertical sections show that they cannot compare in thickness.

Iron Ore south-east of Punxatawney.

But it was especially for iron ore that the examination of the region south-east of Punxatawney was made; and it was, on the whole, unsuccessful. Iron ore deposits were examined, as described above: but in no case could there be found evidence of a regular and persistent bed, excepting the small bed on Wilson's Run, and the ore along the Mahoning creek at Clayville, and that only picked into on the outcrop. Nodular ore masses in shales, no matter how closely they may be packed together at the outcrop where exposed, afford no reasonable basis for iron works. Such deposits frequently yield, and very cheaply, some thousands of tons of excellent iron ore; but they are essentially irregular, changing rapidly, thinning down to nothing without warning, and only regular in this, that they surely bring ultimate loss to those who drive in regular drifts upon them. In working the Clarion Iron Ore, which is a regular plate of carbonate iron ore from six to fifteen inches in thickness, the main entry and workings are in the easily worked slates overlying the ore, the ore bed itself being just above the floor of the mine. These slates hold much kidney ore, and in some places the ore from the slates probably greatly exceeds the amount yielded by the ore bed. But the slates are at times barren of ore for long distances, and not a mine could continue to work the ore at a profit were it not for the small but continuous plate of carbonate (Clarion) ore.

CHAPTER XXIV.

The Fourth or Reynoldsville Gas Coal Basin. Clarion Series or Lower Coal Beds.

The detailed description already given of the mines opened in the Reynoldsville Basin shows that the Freeport Beds are the only ones worked; and that the Lower Freeport Bed yields all the coal shipped to market and nearly all the coal used in the region. It is inevitable that where there is one especially large and good bed in a district, the beds above and below will be entirely neglected. This neglect is not merely in the vicinity of the larger bed, but many miles away it will be difficult to procure accurate measurement of other coal beds in the total absence of worked mines. For local use it is plainly far cheaper for a farmer to buy and haul the small amount of coal needed for his own use from a mine extensively and cheaply worked than to open even the same bed, or still more a smaller and perhaps slightly inferior one, on his own property. Coal is hauled, therefore, from east of Reynoldsville to Luthersburg, a distance of seven or eight miles, in preference to working the three foot beds of that region.

The only mines, therefore, in the Reynoldsville Basin, south of the Sandy Lick creek, opened and worked on the coals below the Lower Freeport Coal Bed, are a few lying on the east and west sides of the basin, along the creek.

The third anticlinal axis, the Boon Mountain axis, crosses the Sandy Lick creek, as already described, near the Clearfield-

Jefferson county line, carrying on its back there the lower beds of the Lower Productive Coal measures. Just east of this axis, and upon it, several coal mines have been opened and worked.

At the village of Rumbarger, Du Bois Station on the Low Grade railroad, a four foot coal is reported as struck in digging wells. It is nowhere to be seen and measured. The main stream of the Sandy Lick creek, and its branches, cut broad valleys, and the hills rise back slowly from the stream on the road to West Liberty, only rising 105 feet in one and one-half miles. The general dip appears to be softly to the south-east from the third anticlinal axis. Going south from Rumbarger to West Liberty, no massive sandstones show on the road, the rocks being grayish slates, usually sandy, and some small thin sandstone pieces. Not a single marked coal bench occurs except the one 65 feet above Sandy Lick creek.

In the railroad cut at Du Bois station, a small 12 inch coal shows, with five feet of fire-clay underlying. Only loose stuff is on top of the coal, the regular roof not being there. This is probably the "4 foot" coal reported as underlying Rumbarger village; and the absence of true roof where the coal shows in the cut, renders it entirely possible that in a normal condition and under cover, the bed may reach three or four feet in thickness. In the railroad cut, one-half mile west of Du Bois station, a massive, fine-grained greenish sandstone shows, passing downward to the south-east and underlying the Rumbarger coal.

The Carrier mine is opened and worked on the Low Grade railroad, three-fourths of a mile west of Du Bois station. The coal as measured in the mine shows (Fig. 104) as follows:

Fig. 104.
Carrier

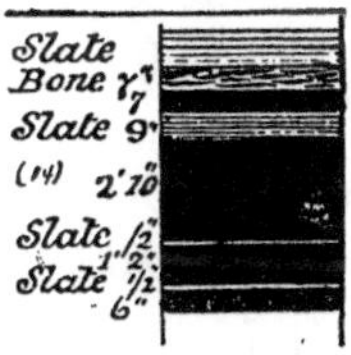

Roof, micaceous sandy slate.	
Bone coal,	0' 7"
Coal,	0 7
Slate,	0 9
Coal,	2 10
Slate,	0 0½
Coal,	1 1
Slate,	0 0½
Coal,	0 6
Fire clay floor.	

The mine runs in north 15° east and fills with water, the coal apparently rising locally to the east and south. The mine is not driven in far, and so far as the coal is exposed it looks well. The hill rises only 20 feet above, made up of thin sandstone. The hills here are low and flat, and the stream valleys broad, cutting out more than one-half of the coal above water level.

Forty-eight feet below this Carrier mine, in the railroad cut, is a 12 inch coal bed, with 12 inches of bone coal and slate on top of it, and thin bedded sandstone for 12 feet over that. Three feet of fire-clay underlie the coal, with gray slates under the clay.

About one mile south of the Carrier and Baum mine are the Rumbarger openings. The lower bed is called a "6 foot" coal, with slate roof; the upper bed, 30 feet above, is called a "2 foot" coal.

About three-fourths of a mile east of Evergreen station on the Low Grade railroad, a railroad cut shows black slates, 10 feet thick, and thin sandstone overlying. The measures are there about flat. Slab Run comes into the Sandy Lick creek, one-half mile east of Evergreen station, making again a broad valley with low hills on each side.

An opening has been made into a coal bed on Fall's creek, about one-half mile above its mouth, and on the east side of the creek. As imperfectly exposed when examined it showed:

Gray slate roof, rotten, . .	4′	0″
Black slate, ...	2	6
Coal, ...	0	7
Soft fire-clay, ...	1	8
Coal, ...	2	0
Fire clay floor.		

The drift was only in about 40 feet when the examination was made, and the coal and the enclosing clays and slates were much decomposed. The fire-clay parting seemed to give promise of thinning down.

Enormous boulders of sandstone, gray and massive, surround the mouth of this drift and lie above it on the hill side for 80 feet. These huge boulders in some cases contain over 3,000 cubic feet of stone. A flat bench, 120 feet above the coal, shows no sandstone, but all gray slates.

These measures are the same as those showing at and around Fuller's mills, on the west side of the Fourth Coal Basin; and the massive sandstone is also brought to daylight along the creek by the subordinate anticlinal axis, west of Reynoldsville.

At Osborne's Mill, one mile up Fall's creek, this massive sandstone shows as a conglomerate rock in places, with numerous small white rounded quartz pebbles imbedded in it.

South of the Sandy Lick creek, opposite the mouth of Fall's creek, and six miles east north-east of Reynoldsville, Bell's Mine is opened and worked. The coal as measured in the mine shows (Fig. 105) as follows:

Fig. 105
Bell's

Roof, black slate, - - - - -	4′	0″
Coal, - - - - - - -	0	6
Thin slate and bone coal, - -	1	3
Black slate, - - - - - -	0	5
Coal, - - - - - - -	4	0
Slate (pyritous), - - - - -	0	2
Coal, - - - - - - -	0	9
Fire-clay floor.		

The main four foot bench of coal is hard and bright, and mines out well. The measures are sinking gently to the south south-east. The mine has never been wrought for shipment, but only in a small way to supply the local demand. It was not being worked at the time of the examination, but an apparently average specimen of the main bench of coal, forwarded to the Laboratory of the Survey at Harrisburg yielded, on analysis, (M'Creath):

"Water, - - - - - - -	0.950
Volatile matter, - - - - - -	32.450
Fixed carbon, - - - - -	59.904
Sulphur, - - - - - - -	1.296
Ash, - - - - - - - -	5.400
	100.000

Coke, per cent., 66.60. Color of Ash, gray.

The coal is bright, compact, contains charcoal and pyrites; also oxide of iron." It forms a coherent coke with a metallic lustre.

The same bed is opened at the Dixon mine, on the south side of the Sandy Lick creek, five miles east north-east of Reynoldsville. The coal as measured in the opening on the east side of the main road shows (Fig. 106) as follows:—

Fig. 106.

Roof, fire-clay and dark clay slate,	1′	2″
Black slate and bony coal,	0	8
Coal,	1	11
Slate, persistent, 0′ 1″ to	0	2
Coal,	2	2
Slate, not persistent,	0	0½
Coal,	2	5
Slate,	0	1
Coal,	0	8
Fire-clay floor.		

The mine is very much troubled by fire-clay "horses," and it is difficult to get any fair average of what amount of coal extended workings will win out; but certainly very much less than the great amount of coal shown in the section of the bed. "Sulphur balls" are numerous in the parting slates.

The second opening on the same bed is on the west side of the main road, the mouth of the drift being only 20 yards distant from mine No. 1.

The coal shows:—

Roof, bony coal and slate.		
Coal,	0′	5″
Fire-clay,	0	1
Coal,	1	2
Slate,	0	1
Coal,	3	3
Slate parting, pyritous,	0	1
Coal,	0	10
Coal in floor,	1	0
Bottom not seen.		

The rocks overlying the coal are soft olive colored shales and gray slates for 120 feet. 70 feet above the coal is a bench on the hill side with a small black slate smut. Some few pieces of a red hematite iron ore of excellent quality, (57 per cent. of metallic iron by analysis, M'Creath,) are found in the shales, 100 feet above the Dixon mine; but the ore evidently exists only in very small quantity, scattered through the shales.

The coal mined from these openings, except where squeezed by the "horses," mines out fairly well, but shows the presence of considerable pyrites.

The London mine is a natural opening in a ravine, about three-fourths of a mile south-west of the Dixon mine, and about on a level with it. It lies on the south of the Sandy Lick creek, about four miles east north-east from Reynoldsville. The coal as measured in the rather imperfect exposure showed (Fig. 107) as follows:

Fig. 107.

London.

Roof, thin bedded black slate,	2′	6″ to 3′	0″
Cannel slate, smooth, and bone coal,	0	8 to 0	9
Coal, good, ...		0	8
Slate, with interleaved coal, . .		7″ to 0	9
Coal, ...		5	0
Floor not seen.			

The coal showed extremely hard and bright even where measured right at the outcrop. The coal mines out in blocks and is excellently fitted to bear railroad transportation.

At a second natural opening on the same bed, one-third of a mile away from the first exposure, the coal shows the same three feet of thin bedded black slates for a roof, and the coal of the same size, hard and bright as before.

A specimen of this coal forwarded to Mr. M'Creath for analysis, yielded:

"Water, - - - - - -	1.150
Volatile matter, - - - - -	27.705
Fixed carbon, - - - - -	65.835
Sulphur, - - - - - -	.930
Ash - - - - - - - -	4.380
	100.000

Coke, per cent., 71.145. Color of Ash, red.

The coal has a dead lustre outside, bright and shining inside, is iridescent, and contains pyrites."

The above analysis represents a very good coal; differing from the coal of the Lower Freeport bed in having decidedly less volatile matter. It is harder also than the Freeport coal.

The size of these openings just described, the London, Dixon, &c., would indicate the presence of the Big Bed, the Lower Freeport coal; for no such large beds are found in the sections made below that coal. But the structure of the region does not warrant the idea of the plunging down of the Lower Freeport coal to the level of these openings; it would require a fault of 80 feet or more to throw it down to these mines, and the exposures of the region around forbid that explanation. Moreover, the coal not merely differs in showing less volatile matter, but the description of the Dixon mine shows that the size of the coal is abnormal, and that it is troubled by horses, pinched down in places, and varies decidedly in measurement in two mines only a few yards apart. All this differs widely from the regularity of the Lower Freeport Coal Bed.

About 90 feet above the London coal is a marked bench, apparently from level and appearance, the bench of the Lower Freeport bed. It is not so plainly and beautifully marked as around Reynoldsville. The hill rises 85 feet above this upper bench, small sandstone pieces showing on the surface, but gives no distinctly marked bench.

Assuming the identification of the bench of the Lower Freeport coal to be correct, the London bed is in the measures 80 to 90 feet below it, making it the Kittanning coal.

Some difficulty was experienced in tracing the Lower Freeport Coal Bed around from where it disappears under heavy cover at the Carrier and Wilson and Sprague mines. The sketch map shows by a dotted line where the outcrop crosses the Luthersburg and Reynoldsville pike, on the eastern limit of the coal, and comes round to join the proved outcrop on Sandy Lick creek. This line necessarily carries it back of and above the London coal.

The Lower Freeport coal crosses the Luthersburg pike just beyond the county line, (Clearfield-Jefferson) a full 100 feet higher than where it goes under the hill at Sprague's mine. Its presence is shown by its bench, not so well defined as farther west, by a line of springs and the smut and clay on the road. At the only place where this bench is at all exposed, near Pheeley's house, the outcrop shows an enormous mass of black slate; and it looks as if, on this extreme eastern edge of the coal bed, much of its coal is replaced by slates, thus explaining the different appearance of the bench.

About 70 feet above this outcrop, just at the Clearfield county line, a small outcrop of slate and a spring indicate the position of the Upper Freeport coal. Gray slates overlie the Upper

Freeport coal for 40 feet, and also make most of the rock between the two coal beds.

One mile west of Evergreen station on the Low Grade railroad, a long cut on the railroad shows:

Small coal.	
Brown shales, with iron ore balls,	5
Black slate, ...	15

The cut is 300 yards long and shows the measures dipping very gently to the north-west, this district being entirely across the anticlinal axis and in the Fourth basin. At this point the hills along the Sandy Lick creek rise very slowly back from the creek, and the small branches also cut out broad valleys.

One and a half miles west of Evergreen station, a small railroad cut shows gray slates, rusty and thin bedded, with a small three inch coal. The measures dip gently to the north-west. One-half a mile further down the creek, black and brown slates and shales take the place of the gray slates and hold a six inch coal. The basin is here almost entirely level, the coal running along nearly flat for a considerable distance.

About one-half a mile east of Pancoast station, on the Low Grade railroad, the hills on the north side of the Sandy Lick creek rise higher and come closer to the stream. Here the Lower Freeport Coal Bed comes into the hills for the first time. It occupies, however, as the map shows, only an inconsiderable area north of the Sandy Lick creek, soon shooting out of the hill top to the northward. Sharp's mine, located at this point, has been already described.

The vertical section of the measures has already been given, (Fig. 65) and described. It will be seen by that section that the London bed at this point is only a two or three foot bed; though it is reported that the outcrop was found on the south side of the Sandy Lick creek, not, indeed, as large as at the London opening, but still as a good, workable bed. This was not seen.

On the south side of the Sandy Lick creek, at Pancoast station, the hills rise high enough to take in the Lower Freeport coal bed, with abundant cover. Pancoast station is the extreme eastern point, therefore, where the Low Grade Railroad strikes the "Big Bed." It follows the outcrop, but about 150 feet below it, to Reynoldsville, where it again leaves it. It has been shown that the mass of this Big Bed coal is south of the Sandy Lick creek.

Passing westward down the Sandy Lick creek, there are no openings on the Big Bed of coal, on the south side of the Sandy, until opposite M'Ghee's mills where the Ohio Company have opened up this Lower Freeport Bed on the outcrop.

The Superintendent, Mr. Brown, reports the trial shaft as showing as follows, (Fig. 108):

Fig. 108.

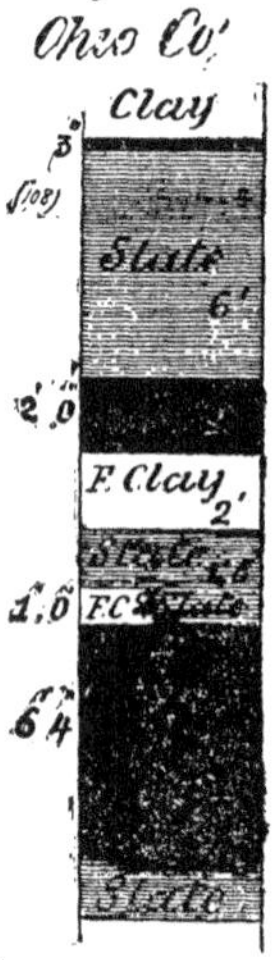

Soil,	4′	0″
Yellow clay,	0	4
Coal smut,	0	3
Fine loose slate,	6	0
Hard coal smut,	2	0
Fire-clay,	2	0
Black slate,	1	6
Whitish clay and slate,	1	0
Coal,	6	4
Black slate in bottom.		

From the above description it appears that the Lower Freeport bed keeps its full size at this point.

The measures at Reynoldsville, on the south side of the Sandy Lick creek, have been already described. The coal beds have been opened for measurement though the Lower Freeport bed is the only one worked. This bed is 150 feet above the Sandy Lick creek.

The coal in the bed of the creek at Reynoldsville had been reported as a large and valuable bed. It was opened for examination by Mr. Jenks, of the Central Land and Mining Company. The coal measured:

Roof, dove colored slate,	4′	0″
Slate,	0	4
Coal,	0′ 6″ to 0	7
Slate,	0	7
Coal,	1	7
Slate,	0 ′2″ to 0	
Fire-clay floor,	4 or more.	

Though this coal is here so mixed with slate as to be entirely worthless, yet it is opened to the north-eastward, where it is well up in the hills as a good sized bed and of excellent quality.

A specimen of this creek coal yielded on analysis (M'Creath):

"Water,	0.800
Volatile matter,	32.020
Fixed carbon,	51.887
Sulphur,	3.593
Ash,	11.700
	100.000

Coke, per cent., 67.18. Color of Ash, gray, with red tinge. The coal is bright, shining, clean, compact, with veins of pyrites and charcoal."

The above analysis represents an utterly worthless coal.

It having been confidently asserted and repeated that the roof slate of this small coal bed had given evidence on several occasions of possessing remarkable heating power, an analysis was made of it by Mr. Ford, at the laboratory of the Survey in Harrisburg, resulting as follows:

"Silica,	48.835
Alumina,	25.235
Protoxide of iron,	2.279
Bisulphide of iron,	.198
Lime,	.165
Magnesia,	.635
Alkalies,	1.780
Sulphuric acid,	.334
Water,	3.090
Organic matter,	17.794
	100.345

The slate is hard, compact, of a very black color, with numerous crystals of pyrites. It is, of course, simply a black slate, and the story of its showing great heating power is clearly a mistake.

For 35 feet above this lower coal bed the measures are black slates chiefly, with nodular carbonate of iron masses scattered through them. But though these ore balls are quite closely packed in places, yet there is no such regularity of deposit as to warrant regular working.

Just on top of these slates, 35 feet above the creek coal, is a coal bed which was opened up on the hill in the village, and measured two feet of coal. It is reported to have been found as a four foot bed at other points, but it certainly only gave two feet in thickness where examined, and the roof and floor seemed smooth and regularly in place.

Twenty-five feet above this coal, 60 feet above the creek coal, and 90 feet below the Lower Freeport coal, is a coal bed, not opened up now for measurement. A row of springs marks its outcrop. It is apparently about a three foot bed, though it is reported as having shown four feet of coal.

It may be noted that the coal outcrop which shows on Fuller's hill, just one mile east of Reynoldsville, 40 feet below the Lower Freeport coal, makes no show on the hillside at Reynoldsville; while the coal 90 feet below the Lower Freeport coal, which shows plainly at Reynoldsville, makes no show on the road side at Fuller's hill though the measures are well exposed at that point.

Westward from Reynoldsville along the railroad, down Sandy Lick creek, a massive fine grained micaceous sandstone, olive colored on the surface in places, shows in the cut, half a mile below the station. The cut shows:

Sandstone, thin bedded, . .	8
Coal, ...	1
Fire-clay and bastard limestone,	6
Massive sandstone in bottom.	

These measures are all rising very decidedly to the subordinate anticlinal axis just west of them, and therefore dip to the southeast, passing underneath the creek coal at Reynoldsville. This cut shows one of those sharp 10 to 15 feet rolls which trouble mining in these lower coal beds.

The steady dip of these measures to the south-east brings up lower rocks constantly on the railroad going westward, and in the cut one-and-a-half miles below Reynoldsville other and still lower coal beds are exposed. The measures show:

Dark colored slates, . . .	8'	0''
Coal, ...	1	0
Fire-clay, ...	4	0
Sandstone and shales, . .	7	0
Black slates, some iron ore balls,	10	0
Coal, ...	2	6
Slates and ore balls, . . .	8	0
Coal, ...	0	6
Impure fire-clay, ...	3	0
Railroad level.		

These measures are dipping decidedly to the south-east.

A little west of this railroad cut, opposite Carrier's saw mill, and on the north side of the Sandy Lick creek, the 30 inch coal of the above section has been opened up about 80 feet above the railroad. The coal measured:

Fire-clay roof, ...	4′
Coal, ...	2
Small slate parting.	
Coal, ...	1
Fire-clay floor.	

The mine runs in north-east and drains. The hill rises 75 feet above the coal, and is covered with thin sandstone pieces of moderate size, to the hill top; and the same sandstone covers the surface of the hillside from the mine down to the railroad.

A railroad cut at this point affords a cross section which shows perfectly a horse of clay in coal, and also the forking of a coal bed, thus:

Fig. 109.

Horse of clay in Coal bed, exposed in cutting of Bennett's Branch R.R., near Carriers Saw Mill; Jefferson Co. Pa: Report of F. Platt, 1874.

At the mouth of Deemer's Run, Carrier's Station on the Low Grade railroad, two miles below Reynoldsville, the railroad cut shows 20 to 25 feet of massive sandstone, mostly whitish, with some brownish colored layers. The measures are still sinking to the south-east, but this is about the centre of the arch of the subordinate anticlinal axis; and the rocks exposed belong to the Seral Conglomerate at the base of the Lower Productive Coal Measures and the rocks immediately below it.

Capt. J. M. Steck, of Brookville, Jefferson county, kindly furnishes the following record of an oil boring put down on the bank of the Sandy Lick Creek at this point:

		Feet.		Total depth.
1. Wooden pipe, (unknown,)	- - - -	25		25
2. White sand-rock, XII *Seral,*	- - - -	35		60
3. Shale rock,	XI *Umbral,*	32	204′	92
4. Red rock,		39		131
5. Slate rock,		33		164
6. Red rock,		16		180

		Feet.		Total depth
7. Slate rock,	XI *Umbral,*	19	204′	199
8. Slate rock,		3		202
9. Slate rock,		28		230
10. Blue sand-rock,		15		245
11. Slate rock,		15		260
12. Red rock,		4		264
13. Blue sand-rock,		56	(soot at 322)	320
14. Slate rock,		10		330
15. White sand-rock,	X *Vespertine,*	26		356
16. Slate rock,		54		410
17. Fine sharp sand-rock,		9		419

Salt water at 419 feet. Strong gas at 395 feet.

White sandstone shows on the hillside up Deemer's Run, far above the railroad. Messrs. Young and Fagen made a section up the hill at the head of the run. It is going with the rise of the rocks, and therefore exaggerates their thickness.

Fig. 110. The hillside shows (Fig. 110) as follows:

Hill top.	
Shale,	30
Shaly sandstone,	20
Massive white sandstone,	35
Shale,	70
White sandstone,	110
Sandy Lick Creek.	

The centre of the subordinate anticlinal axis having been passed near Deemer's run, the measures dip slightly to the north-west. But this dip is not continuous across the basin to meet the south-east dip of the rocks from the Fourth anticlinal axis; but rather a series of small flexures, from this subordinate anticlinal westward to the Fourth anticlinal axis. These flexures are not great, and their effect is to keep the railroad running for the whole distance from Deemer's run to Fuller's mills, in about the same rocks; the Seral Conglomerate making the country rock all along the bed of the creek, the rolls sometimes carrying it above the railroad slightly, and sometimes the railroad running just on top of it, and coal opened within 50 feet of the level of the track. The effect of this, of course, is to confine the coal measures in this part of the Fourth Coal Basin to the coal beds immediately above the Seral conglomerate, and below the Lower Freeport Coal.

Rinestine's opening is up Six Mile Run, about one-half mile above its mouth, four miles west of Reynoldsville, and 90 feet above the Sandy Lick creek. It shows:

Roof, slate, with many iron ore balls.	
Bony coal,	3′ 0″
Coal,	3 3
Fire-clay floor.	

The coal is only opened up for say 50 feet in from the outcrop, but shows hard and firm, with only one small one-half inch slate layer in the three foot three inch bench. It dips gently to the east 30° south. Massive sandstones fill the valley of Six Mile Run below the coal. The coal occupies about the same place as the Fuller's Mills coal farther west, and they are probably the same bed. The opening opposite Carrier Station, already described, is also most probably on the same bed.

Ammerman's Mine is opened and worked about five miles west of Reynoldsville, on the south side of the Sandy Lick creek, and 75 feet above the railroad. The coal measured in the mine:

Clay slate rock above,	
Roof, hard black slate,	2′ 0″
Coal,	3 8
Black clay slate parting,	5″ to 0 7
Coal,	8 to 1 0
Clay slate floor.	

The mine is not yet driven in far, but the coal is black and firm, columnar in structure, and carries much iron pyrites. The mine runs in south 70° west, and rises. One small roll shows already in the mine.

The Rocky Bend coal company have opened and are working a mine near Fuller's Mills station, on the Low Grade railroad, six miles west of Reynoldsville. The mine is opened on the north side of the Sandy Lick creek, about 50 feet above it. The coal as measured in the mine shows (Fig. 111) as follows:

Fig. 111.

Rocky Bend

Roof, black slate.	
Coal,	1′ 6″
Slate,	0 2
Coal,	2 2
Cannel coal,	0 6
Slate,	0 4
Coal,	2 10
Fire-clay and clay slate floor.	

The mine runs in east 25° north, and the rooms are turned

off south 15° west to the rise of the coal. Black, firm, bright coal was struck not far in from the outcrop; but there appears evidence of a thinning down of the coal from the thickness given in the section above.

The Superintendent reports that the Buffalo Gas works returned him the following description of the gas power of the coal to the ton:

Gas,	9,000 cubic feet.
Coke,	37 bushels
Candle power,	13.6.

As this coal bed from its size and character gives importance to this region, it should be added, as evidence of irregularity to be expected in working, that three openings had previously been made between this Rocky Bend mine and Fuller's mill about 600 yards west of it.

In the first opening, which ran in north 60° east, the coal measured:

Roof, rotten clay.		
Bony coal,	2′	2″
Coal,	2	1
Clay slate parting,	0	7
Coal,	2	11
Fire-clay and clay slate floor.		

The drift was not in entirely past the outcrop coal; and the upper bench of coal showed harder and better than the lower. The coal showed a tendency throughout to take on a slaty cannel appearance, and to run into an ashy coal.

The second trial opening ran in north 20° east and found only loose stuff for ten yards in, and then only a small streak of coal in clay. The floor rises very rapidly, 10 feet in about 70. Where measured about 25 yards in from the mouth of the drift, the coal shows:

Coal in roof.		
Coal,	5′	0″
Fire-clay,	0	6
Coal,	2	6
Fire-clay floor.		

The bottom bench yields harder and better coal than the upper; and the faces of the coal were covered with infiltrated clay.

The upper bench is slaty, not so much in regular slate partings which can be picked out, but in taking a slaty and bony character, at times for almost the whole upper bench.

At the third opening, which runs in about north, the coal shows:

Coal in roof.		
Coal, soft, ...	2′	3″
Slate and fire-clay, . . .	1	1
Coal, ...	2	4
Fire-clay floor, soft.		

The coal was still near the outcrop and much weathered; the lower bench giving the best show of coal farther in.

The coal along the face of the hill seems somewhat disturbed, and some trial pits on the level west of this opening, failed to find the coal. As the bed is rising to the Fourth anticlinal axis, they were probably below the coal.

A vertical section made up the hillside, north-west of the *Fig.* 112. Rocky Bend mine shows as follows, (Fig. 112):

Hill top.	
Sandstones, ...	50′
Grey slates, ...	42
Bench.	
Grayish slates, ...	38
Bench.	
Sandstone, thin bedded, with grayish slates, and an occasional massive layer of sandstone, ...	105
Slight bench.	
Concealed measures, ...	30
Small black slate outcrop.	
Concealed measures, ...	33
Coal, Rocky Bend, ...	7
Sandstone, ...	45
Level of Sandy Lick creek with small coal in creek bottom.	

The hill is narrow and steep, lying in the ox-bow in the creek.

The Seral conglomerate is the bottom of the above section, the small coal in the creek, belonging apparently in it. There are here, therefore, three hundred feet of measures above the Rocky Bend coal and the Lower Freeport coal not reached. This agrees with the sections which will be given in the report upon the north-east part of the Reynoldsville Basin.

The coal found in the hillside west of Fuller's mills is probably the same Rocky Bend coal; the lowest of the workable coals of the Lower Productive Coal Measures.

The Sandy Lick creek, west of Fuller's mills, runs through the Seral conglomerate for some distance; but it soon rises above the creek level and at Iowa Mills makes the sides and finally the crest of the hills. The centre of the Fourth anticlinal is at or about Port Barnet, two-and-a-half miles east of Brookville; and the creek for some miles has for its country rock the measures of X, formation XI (*Umbral* red shales) being here very much thinned down.

CHAPTER XXV.

Coals and Ores of the Reynoldsville Basin.

For convenience in reference, the analyses of the coals given in the above detailed report are grouped together in the table below; the number of the figure of the section in the report being added to each name of a colliery:

		Water.	Volatile matter.	Fixed carbon.	Sulphur.	Ash.	Coke, per cent.
(73)	Seley, Upper	0.85	31.20	59.882	1.368	6.70	67.950
	Middle	1.04	31.61	62.464	.736	4.15	67.350
	Lower	0.96	32.32	58.640	1.230	6.85	66.720
(71)	Sheesley, Upper	1.60	30.70	63.791	.639	3.27	67.700
	Lower	1.48	29.22	65.022	.608	3.67	69.300
(68)	Diamond, Upper, } *Not*	1.10	29.99	46.639	3.101	19.17	68.910
	Middle } *worked.*	1.19	32.81	55.316	2.284	8.40	66.000
	Lower Bench, Middle	0.95	35.13	59.304	1.436	3.18	63.920
	Lower	1.12	33.86	60.692	1.278	3.05	65.020
(69)	Hoover, Upper	0.96	32.68	59.097	1.063	6.20	66.360
	Middle	1.10	30.80	62.524	.776	4.80	68.100
	Lower	1.10	32.90	62.174	.726	3.10	66.000
(74)	Sprague	1.43	31.94	62.109	.531	3.99	66.630
(84)	Wachob	1.30	32.57	62.567	1.023	2.54	66.130
(89)	J. Thomas	0.95	31.59	60.520	1.440	5.50	67.460
(102)	Anthony	0.95	35.87	58.218	2.302	2.66	63 180
(87)	P. Hawk	0.95	33.55	60.523	1.167	3.81	65.500
(100)	Ruth	1.06	34.14	61.172	.678	2.95	64.800
(94)	Pantall	1.10	31.17	63.544	1.016	3.17	67.730
(95)	M'Kee	1.05	33.15	58.405	1.295	6.10	65.800
(90)	Weaver	1.00	33.20	59.428	2.042	4.33	65.800
(96)	Wingert	1.15	32.07	60.428	1.702	4.65	66.780
(98)	Hum, Upper	0.92	35.44	59.962	.848	2.83	63.640
	Lower	1.00	33.26	63.081	1.139	1.52	65.740
(70)	W. Reynolds	1.44	32.46	63.011	.639	2.45	66.100
(66)	Sharp, Upper	1.32	31.44	62.578	.892	3.77	67.240
	Lower	1.57	33.43	61.285	1.055	2.66	65.000
(107)	London	1.15	27.705	65.835	.930	4.38	71.145
(76)	Strouse	1.30	30.22	63.617	.763	4.10	68.480
	Creek, Reynoldsville	0.80	32.02	51.887	3.593	11.70	67.180
(105)	Bell	0.95	32.45	59.904	1.296	5.40	66.600
	Graff	1.05	29.73	59.781	1.389	8.05	69.220
(103)	M'Farland	1.02	30.19	57.943	2.757	8.09	68.790

The above analyses indicate very great regularity in the character of the coal. The volatile matters range from 30 to 35 per cent., and average about 32 or 33 per cent.

Mr. Sharp furnishes the following record of a test of the coal from the Diamond colliery, at Reynoldsville:

"Tested at Buffalo Gas Works, and reported by the Superintendent.

Gas 9,250 feet to one ton, (2,000 pounds of coal.) 14.3 candle power.

Thirty-nine bushels of coke from 2,000 pounds of coal. The coke good, with very little clinker."

It is noteworthy that all analyses made of coals in this basin indicate a gas coal character; not only the coal from the Freeport beds, but from those lying below, while in the First Coal Basin, analyses of coal beds A, B, D and E, from the conglomerate up to and including the Upper Freeport Coal Beds, all indicate a steam coal, the volatile matters ranging from 19 to 22 per cent.

In view of the use of the coal in the future with iron works, it is of importance to note that Mr. M'Creath's analyses of two of these coals for phosphoric acid show it to exist in very small quantities, thus:

	Per ct. in Coke.	Per ct. in Ash.
Diamond mine, ...	Trace.	Trace.
Hoover's mine, upper, ...	.071	1.145
Do. middle, ...	.008	.166

This upper bench of Hoover mine is the top part of the main bench, not now worked.

The greater part of all the coal now shipped to market, is sold for gas making purposes to northern New York and Canada; though the coal makes a strong burning steam coal and has also a market for that purpose. The line of shipment to north-western New York State is by the Low Grade railroad east to Driftwood, then by the Philadelphia and Erie railroad to Emporium, thence by the Buffalo, New York and Emporium railroad to Buffalo and Canada. At present, there is no good gas coal mined along the route, and there is therefore a large and growing market awaiting this coal, in reaching which market it has the advantage in distance over its competitors. It could also reach Philadelphia and New York with about the same distance by railroad as the Westmoreland gas coal.

It is difficult to make an estimate of the amount of the Lower

Freeport Coal Bed remaining in the basin between the Sandy Lick creek and the Mahoning creek. The lines of outcrop on the accompanying map shows the general outlines of the bed. It seems not too high an estimate to say that of the nominal average, broadly taken by square miles, one-half must be given for outcrop coal, lost on ravines, with insufficient cover, thinning of the bed, and inaccessibility. But as every square mile of a five foot coal bed contains 5,000,000 tons of coal, there is an enormous amount of coal to be yielded by this region.

The regions lying up Soldier run and Trout run are easily accessible by railroad, but it is yet to be decided whether the great quantities of fine coal north-west of Punxatawney will be reached by a branch railroad from the Low Grade railroad, or ultimately go down Mahoning creek; but as their natural market is at present to the northward, the coals will probably come out to the Low Grade railroad.

Iron Ores of the Basin..

There is in all the vertical sections given in this basin, a striking lack of iron ore. The only ore deposits of any considerable size examined and described, are the moderate sized deposits on Hog Shanty run, the ore bed on the Mahoning creek at Clayville, and several deposits of balls of carbonate iron ore in shales, not sufficiently regular to be of workable value. It is said that a large iron ore bed was opened many years ago, south south-west of Fuller's mills about five miles, and that an analysis of it by Prof. F. A. Genth, of Philadelphia, showed it to be of excellent quality and rich in metallic iron. It can not now be found by those who opened it many years ago, and therefore was not examined. In the position of the Clarion iron ore, one exposure, already noted, about 150 feet below the Lower Freeport coal bed, the only ore shown consisted of numerous ore balls on the shales, but no regular ore bed.

The Low Grade railroad has only recently been opened through this region, and before having railroad communication there was but little inducement to open up iron ores. And the great depression in the iron business operates now to prevent any thorough examination for iron ores. But there are numerous excellent sites for furnaces at and around Reynoldsville; there is an abundance of excellent coking coal, and the coke can be cheaply laid down at the furnace stack; the deposit of limestone below the Lower Freeport coal bed is developed in size at Reynoldsville, (it is lacking usually through the basin,) and it is probable, from the ore beds seen partially opened, and from the statements made, that a revival of the iron trade will develop workable iron ore beds, and make this an iron smelting region.

CHAPTER XXVI.

The Fourth or Reynoldsville Basin, continued North-eastward to the Little Toby.

The second part of the Reynoldsville Gas Coal Basin lies between the Sandy Lick creek on the south-east and the Little Toby creek on the north-east. It will be much more briefly described than the first part of the basin, in chapter XXIII, for it lacks the presence of the valuable Lower Freeport Coal Bed, which is so prominent a feature south of the Sandy Lick creek, and, moreover, much of this part of the basin is at present entirely without access to market.

The geology of the basin has already been completely described. The basin is bounded on the east by the Third anticlinal axis, Boon's Mountain; on the west by the Fourth anticlinal axis, about the line of Mill creek; and in the centre the measures are thrown up by the subordinate anticlinal axis which crosses the Sandy Lick creek west of Reynoldsville. It has been stated that the only exposure of the Lower Freeport Coal Bed north of the Sandy Lick creek is the small area just north of Pancoast Station, on the Low Grade railroad. The coal beds worked, therefore, in this second part of the basin are those below the Lower Freeport; the same beds which are so little opened in the part of the basin between the Sandy Lick and the Mahoning creeks. By putting these developments together we have a tolerably well opened section of the measures from the Seral Conglomerate up to the Mahoning sandstone.

Going north-west from Reynoldsville, the measures are rising rapidly to the subordinate anticlinal axis; and the measures below the Lower Freeport are seen high up on the hill, which rises back slowly to the north-west, 400 feet high. Nothing new is developed, and the openings made on the coals, 50 and 165 feet above the creek, respectively, only show in the lower bed a 20-inch coal and in the upper a 15-inch coal. North-west towards Brookville, along the pike, these small coal-shows are found in the sides of the valley of Prior Run, near the school house; and in the valley to the north-west.

A vertical section made up the hill north-west from M'Ghee's Station, on the Low Grade railroad, finds the measures rising rapidly to the north-west and the hill sloping gently back in the same direction. The show is the same as on the hill north-west of Reynoldsville.

The coal bed on the north side of the Sandy Lick creek, 110 feet above the creek, abreast of M'Ghee's Mills, was opened up on the outcrop, but proved to be an average two foot bed and was abandoned.

The P. Cox mine is opened, and has been worked, on the north side of the Sandy Lick creek, one and a half miles north-west of M'Ghee's mills, and about 230 feet above the creek.

The coal, as measured near the mouth of the mine, shows, (Fig. 113,) as follows:

Fig. 113.

Roof, fine grained sandstone,		3'
Kidney ore in fire-clay,	0' 4" to 0 6"	
Coal,		1 4
Slate, persistent,	0 1 to 0 2	
Coal,		2
Floor not seen.		

The coal dips east and south. It is reported as having yielded when worked a full average four feet of good coal.

The coal is hard and good, even near the mine mouth where measured.

The iron-ore ball deposit resting upon the coal is very uneven in thickness, sometimes showing several inches of closely packed nodules of ore, and again not more than an inch of fire-clay, entirely without ore, separating the coal and the sandstone.

An opening on Painter run, on the bed which lies 62 feet below the P. Cox coal, only showed 22 inches of coal, with much slate in that. This is the same bed as the "2 foot" bed opened on the north side of the Sandy Lick creek, abreast of M'Ghee's mills, 100 feet above the stream.

Fig. 114. *Falls Creek.*

Where the main road from Reynoldsville to Brockwayville crosses Fall's creek, the following section is seen, (Fig. 114.)

Hill top.	
Thin gray slates, ...	20′
Small coal.	
Slates, ...	20
Small coal.	
Slates, ...	50
Bastard sandy limestone.	
Sandstone and concealed measures, . .	45
Black slate outrcop.	
Shales and slates	55
Fire-clay.	
Slates and shales, ...	60
Level of Falls' creek.	

On Kyle's property, near this point, a coal has been opened 43 feet above Fall's creek. It shows now two to two and a half feet of coal, but is called a "five and a half foot bed."

Eighty-two feet above the coal, and 125 feet above Fall's creek, a bed of dark blue limestone is opened up, two feet thick, with black slate and coal smut underlying.

Coal has been opened by Mr. Hunter, one mile south of Rockdale. The coal is a "3 foot" bed, now fallen shut and not to be measured; the roof is black slate. The measures overlying are grayish slates, and 70 feet above the coal there is considerable lean brown hematite iron ore scattered through the slates; and also a show of bastard limestone.

On the J. Crawford place, one and a half miles south-west of Rockdale, an outcrop opening on a coal bed just a few feet above the stream shows:

Thin shaly sandstone overlying.	
Roof, black slate, ...	3′
Coal, ...	2 to 3
Fire-clay floor.	

Massive sandstone pieces lie on the hill 110 feet above, and then 30 feet of shales to the hill top; but no marked bench exists in all this distance. At another point, 90 feet above this opening, on the same property, a shaft has found a small 18 inch coal with a fire-clay floor and shaly slates on top.

At Rockdale, 60 feet above the level of Fall's creek, coal has been struck in wells. Sandstone overlies the coal according to the statements made, and certainly huge masses of sandstone cover the hillside from the creek to the hill top, 15 feet above the coal. Mr. S. Crawford reports the coal as having shown, (it has been shut for 15 years):

Coal, ...	3′	
Cannel coal, ...	1	6
Coal, ...	1	6

The description probably exaggerates materially the size of the coal.

The subordinate anticlinal has here brought up very high the massive sandstones of and above the Seral conglomerate; and the coal is probably the Fuller's Mills coal, with which the above description would somewhat correspond.

Going north from Rockdale the sandstone boulders end at 67 feet above the creek, and shales come in for 30 feet on top. At the school house, at the road forks, there is a small black slate show, 107 feet above Fall's creek, and over that 15 feet of soft shales to hill top.

On Dougherty's place, one and a quarter miles north of Rockdale, two coals have been opened up. The lower coal bed, which is opened almost in the bed of a small stream, and was covered with water when examined, gives:

Sandstone roof.		
Bone coal and black slate, ...	0′	9″
Coal, showing, ...	1	0
Bottom not seen.		

Though only 12 inches of coal show it is evidently a much larger bed. It is called a "5 foot bed;" and is reported a strong steam coal.

The opening on the upper bed, 40 feet higher, is now fallen shut and the coal could not be measured. The rocks overlying are sandy grayish green slates. 50 feet above this old mine is a bench on the hillside, not opened up.

At Davenport's mine, only a short distance away, the upper of Dougherty's coal beds was formerly opened up. It is now nearly fallen shut, but would indicate about two and a half to three feet of coal, with four feet of black slate overlying.

About one and a half miles north-east of Rockdale, Mr. A. M'Cullough has partially opened up the outcrops of his coal beds.

The following vertical section was compiled at M'Cullough's chiefly from imperfect exposures, (Fig. 115):

Fig. 115.

Hill top.		
Slates,	85′	
Bench of coal.		
Shales and slates,	50	
Coal,	3	6″
Shales and gray slates,	50	
Coal.		
Concealed measures,	9	
Limestone,	3	
Level of creek.		

Where the limestone was partially opened it was dark colored, with considerable iron. In two or three other places where opened it carried a considerable, but varying quantity of carbonate iron ore on it.

The coal over the limestone was not opened up. It is in the geological position of a good bed of coal, and one quite extensively opened and worked to the north-west, as will be seen from the description to follow.

The middle coal, 50 to 60 feet above the limestone, is opened and worked. The coal, as measured in the mine, shows (Fig. 116) as follows:

Fig. 116.

Roof, black slate,		4	
Coal,		2	
Slate, pyritous,		0	0½′
Coal,	1′ 2″ to	1	6
Slate,		0	1
Coal,		0	2
Fire-clay floor, hard.			

The coal is firm and hard, and mines out well. A specimen of the coal yielded, on analysis, (M'Creath):

"Water,	1.320
Volatile matter,	33.920
Fixed carbon,	53.905
Sulphur,	1.505
Ash,	9.350
	100.000

Coke, per cent., 64.76. Color of Ash, gray with pink tinge. The coal has a dull lustre, is somewhat slaty, containing much iron pyrites."

The upper bed is not opened up. It is called a "3 foot bed" from old openings; and lies 50 feet above the middle bed.

The upper and middle bed here represent the two beds at Dougherty's; and the three beds of the section (Fig.115) are the coal beds of this whole region passing into and under almost all the hills, and above water level.

At Cooper's place, north of M'Cullough's, the opening shows:

Thin sandstones, with slates.		
Black slate,	4′	
Bone coal,	1	
Coal.	3	6′
Fire-clay floor.		

Apparently the middle bed of the three beds named above.

H. H. Smith's mine is opened and worked one mile north of Rockdale.

The coal as measured in the mine lies (Fig. 117) as follows:

Fig. 117.

Greenish gray slates,	5′	0″
Black slate,	2	6
Coal,	4	8
Fire-clay floor.		

The mine was partially filled with water when examined, but the above description is not far from correct.

The measures for 65 feet above the coal are thin slates and shales.

West of this mine, at W. Smith's, the upper coal bed, "three feet thick," is struck in the well of the house, the distance between the beds being 60 feet.

At W. Patten's Mine, two miles north-west of Rockdale, coal is opened and worked. The coal as measured in the mine parts (Fig. 118) as follows:

Fig. 118.

Tough gray slates,	4′	0″
Black slate,	1	0
Coal,	2	0
Slate,	0	1
Coal,	2	11
Fire-clay floor.		

The coal was measured near the outcrop, the mine being partially filled with water; but the coal looks good and hard and mines out well.

Going north over the hill, the same coal bed is partially opened up in a ravine. The black slate roof is found here slightly thinner, and in one of the old openings limestone and some iron ore were exposed under the coal many years ago. A bench shows markedly on the hill side 60 feet above this coal. The hill rises 85 feet above this bench, or 145 feet above the coal, and is made up of soft gray slates and shales, with no sandstone, except some small pieces near the hill top.

Samuel Patten's mine, near the W. Patten mine, exhibits coal from three and a half to four feet thick, with black slate roof. The mine runs in north and drains imperfectly. The floor could not be seen.

The coal mines out well, good and bright. A specimen of the S. Patten coal yielded on analysis, (M'Creath):

Water,	1.870
Volatile matter,	32.450
Fixed carbon,	61.103
Sulphur,	.547
Ash,	4.030
	100.000

Coke, per cent., 65.68. Color of Ash, cream.

The coal has a dull lustre, is very friable, iridescent, with a small amount of pyrites.

Fig. 119. Rattle Snake Ck.

Fig. 120. Brockwayville

The hill rises high above this coal bed, very prettily terraced, all made up of soft shales and slates.

A vertical section made at J. J. Stewart's on Rattlesnake Run, two miles northwest of Rockdale (Fig. 119) shows these three coal beds of the region and the measures for about 100 feet above; the four foot bed, with the limestone underlying, making nearly the bottom of the section.

In another vertical section made at Brockwayville, on Little Toby creek, (Fig. 120) this same four foot coal bed is nearly at the top of the section;

and the Seral Conglomerate lies not far below the bottom of the exposed section. By putting these two sections together the whole series of coal measures occupying this Fourth Coal Basin, below the Lower Freeport Coal bed, is given in one section.

The five-foot coal at the bottom of Fig. 119 and the four-foot coal near the top of Fig. 120 are the same coal bed.

The section at Rattlesnake Run reads (Fig. 119) as follows:

Hill top.	
Shales and slates, - - - - -	75′
Limestone, bastard, sandy, - -	3
Shales and slates, - - - - -	42
Coal. - - - - - - - - -	2
Slates, - - - - - - - - -	62
Coal, reported as - - - - -	3
Gray slates and sandy shales, - -	58
Black slate, - - - - - - - -	10
Coal, - - - - - - - - -	4
Level of Rattlesnake Run.	
Limestone and fire-clay, - - -	5
Concealed measures, - - - - -	30
Coal, - - - - - - - - -	2

On this section the small coal below the "five foot bed," at the bottom of the section, is compiled from the valley to the south, where the stream cuts deeper; and the limestone underlying the coal is not opened up for exact measurement.

One of the benches on Rattlesnake hill, above the bastard limestone, may represent the Freeport Coals just caught in this one high hill; but the benches have never been opened and only occur on this high knob.

Stewart's Coal Mine is opened and worked just above the level of Rattlesnake Run. The coal as measured in the mine reads (Fig. 121) as follows:

Fig. 121.

Stewart's (121)

Black Slate 10

Slate 6″

2′3″

Bone 1″

1′ 0″

Slate 1″

10″

Fireclay

Roof, black slate, - - - -	10′	0′
Cannel slate, smooth, - - -	0	6
Bone coal, - - - - - -	0	1
Coal, - - - - - - - -	1	0
Slate parting, - - - - - -	0	1″
Coal, - - - - - - -	9″ to 0	11
Fire-clay floor, hard.		

The mine runs in north-west and drains; and the coal mines out very handsomely.

A specimen of the Stewart coal yielded on analysis (M'Creath):

"Water, - - - -	1.830

Volatile matter, - -	34.270
Fixed carbon, - -	58.353
Sulphur, - - -	.767
Ash, - - - -	4.780
	100.000

Coke, per cent., 63.90. Color of Ash, gray.

The coal has a glossy lustre, is compact, and contains a small amount of iron pyrites."

The above analysis represents an excellent coal; rich in vola tile matters, and with a low percentage of sulphur and ash.

The coal 68 feet above this Stewart coal is not now opened so as to be examined; nor the coal 130 feet above, called a "two-foot bed." The bastard limestone above is very sandy and appeared to be entirely worthless.

Dennison's Mine is opened on the same coal bed as Stewart's, and about one mile north-west of it. The coal shows thus:—

Roof, black slate.		
Cannel slate, smooth, . .	6″ to 0	8″
Coal, ...	2	0
Bone coal, ...	0	1
Coal, ...	1	0½
Floor not seen.		

The coal mines out bright and clean, with an occasional small show of iron pyrites.

This mine lying north-west of Stewart's is necessarily much higher, as the coal is rising decidedly to the north-west. A small bench runs along the hill side about 50 feet above Dennison's mine. It has never been opened.

The coal which is at Stewart's, 30 feet below the "four-foot bed," is opened at Dennison's, 30 feet below his coal. It shows two feet of coal, with black slate roof, and over this gray greenish slates, with some scattered pieces of lean sandy hematite iron ore.

A. Key's mine is opened on this same four foot bed one-half mile east of Dennison's and lies thus:

Roof, black slate.	8′	
Cannel slate, ...	0	6′
Coal, ...	3	6 or more.
Floor not seen.		

The mine is now filled with water and the full thickness and bottom could not be obtained.

Seven feet below this coal a three foot bed of limestone has been opened in the run. The identification of these openings of Key's, Stewart's, Dennison's and M'Cullough's is complete.

The vertical section at Key's mine reads as follows:

Hill top.	
Interval not measured.	
Bench, coal.	
Concealed measures, ...	80′
Bench.	
Concealed measures, ...	50
Bench,	
Concealed measures, ...	60
Coal, Key's, ...	4
Limestone in creek bed.	4

The enclosing measures are apparently gray slates and shales. This section agrees with these already given at Stewart's and M'Cullough's.

The E. Calhoun mine is opened on the four foot bed just south-east of Key's. The coal shows there:

Roof, black slate.	
Cannel slate.	0′ 6′′
Coal, ...	3′ to 4
Floor not seen.	

At W. J. Calhoun's, a natural opening in a ravine shows the four foot coal bed and the small bed below.

The lower bed looks good, hard and bright, but apparently does not exceed two feet in thickness.

The four foot bed, 32 feet above this lower bed, has been worked on the outcrop, thus:

Roof, black slate.		
Cannel slate, ...	0′	8′′
Coal, ...	3	2
Bottom not seen.		

The coal is bright and hard, and mines out well, resembling exactly in general appearance the coal produced from the other mines on this bed already described.

A specimen of the coal from this W. J. Calhoun opening yielded, on analysis, (M'Creath):

"Water, - - - - - -	1.200
Volatile matter, - - - - - -	33.630
Fixed carbon, - - - - - -	55.796
Sulphur, - - - - - -	1.504
Ash, - - - - - - - -	7.870
	100.000

Coke, per cent., 65.17. Color of Ash, gray.

The coal is bright, shining, rather compact, with slate and iron pyrites."

This specimen was taken from near the outcrop, and probably does not do justice to the general character of the coal. It has, however, a gas coal character.

Fifty feet above this W. J. Calhoun coal bed is an outcrop of black slate on a bench not opened up. The measures are the same as already given in the Rattlesnake section, (Fig. 119.)

Coal is opened at Morrison's, three miles north north-east of Rockdale. It is an imperfect opening, and shows:

Gray slate on top.
Roof, black slate.
Coal, ... 2′ 6″ to 3′
Floor not seen.

At Frost's place, one mile south of Brockwayville, the following vertical section was made, (Fig. 122):

Fig. 122.

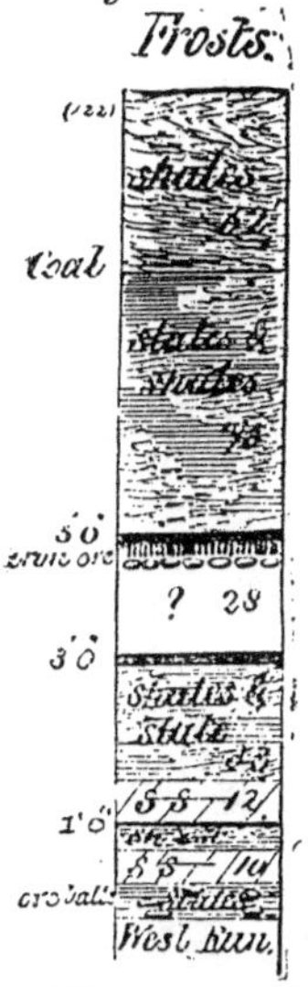

Hill top.	
Shales,	52′
Coal.	
Sandy gray slates and shales,	75
Coal, reported as,	3
Limestone, with iron ore underlying,	4
Concealed measures,	28
Coal, reported as,	3
Shales and slates,	33
Micaceous sandstone,	12
Coal, small,	1
Shales and slates,	7
Brown sandstone,	10
Thin slates, with iron ore balls,	10
Level of West Run.	

Of the coal beds in this section, none are now worked. The "three foot" bed below the limestone, (30 feet,) is opened enough to prove the three feet of thickness, but not to allow an accurate measurement of the bed.

The quantity of blue carbonate iron ore in the shales, just above the level of West creek, is very great on the outcrop, but it has never been opened out to see if a regular bed exists there.

South of Brockwayville, at Dr. Clark's bank, 70 feet above the Toby creek, a coal has been opened up on the outcrop in the following section:—

Sandstone overlying.
Black slate roof.
Coal, ... 3′ 4′
Fire-clay floor.

J. Keys' mine is opened and worked one-half mile west of Brockwayville. The coal holds as measured in the mine, (Fig. 123,) as follows:

Fig. 123.

Roof, black slate,	10′	
Cannel slate,	0	6″
Coal,	3	6
Slate,		0½
Coal,		6
Fire-clay floor.		

The coal mines out well, but holds considerable iron pyrites.

A specimen of the J. Keys coal yielded on analysis, (M'Creath):

Water,	1.360
Volatile matter,	38.720
Fixed carbon,	53.683
Sulphur,	2.047
Ash,	4.190
	100.000

Coke, per cent., 59.92. Color of Ash, red.

The coal has a glossy lustre, is rather compact, containing a small amount of iron pyrites.

The above analysis exhibits a coal unusually rich in volatile matter, but carrying too much sulphur.

The appearance of this coal, the roof, floor, cannel slate, &c., all render it nearly certain that this is the same coal bed opened at Stewart's on Rattlesnake Run, Calhoun, &c.

The vertical section at this point, west of Brockwayville, (Fig. 120) reads as follows: (the cut of the section is (with Fig. 119) on page 214, preceding:

Hill top.	
Shales and slates,	45′
Sandstone,	23
Shales and slates,	32
Coal, Keys',	4
Gray sandy slates,	30
Coal.	
Shales and slates,	85
Coal, small.	
Shales and slates,	45
Coal, not seen, reported as	4
Concealed measures,	10
Level of Toby creek.	

The coal bed lying 30 feet below Keys is opened at M'-Knights, but only shows two and a half feet of coal, rising to three feet in places, with a black slate roof. Floor not seen.

The "four foot" coal reported in the wells in the village of Brockwayville could in no place be seen and measured.

To the east and south-east of Brockwayville there are not many coal openings; but about two miles east of the village, and on the north side of the Little Toby creek, the Steven's mine is opened and shows:

Black slate roof.	
Coal, ...	3' 4" to 3' 6"
Fire-clay floor.	

The mine runs in to the north-west and drains, the coal apparently dipping to the south or south-west.

Meeker's mine, south of Brockwayville, has:

Sandstone on top.		
Slaty sandstone, ...	0	6
Sandstone, massive, . . .	2	0
Black slate, ...	2	0
Coal, ...	2	0
Bottom of coal not seen.		

The mine is fallen shut, and cannot be accurately measured.

Vince's mine, three-fourths of a mile south-west of Brockwayville, shows:

Roof black slate.	
Coal, ...	3' 4"
Fire-clay floor.	

The coal dips to the south-east.

Coal has also been opened on Rattlesnake run, not far from its mouth, south of the Little Toby creek, at Lane's mills.

From here the Fourth Coal Basin continues on to the north-east, to and beyond the Philadelphia and Erie railroad, bounded on the east by the very decided Third anticlinal axis of Boon's mountain, and on the west by the Fourth anticlinal axis which crosses the Little Toby creek, just west of Galusha's mills, three miles north-west of Brockwayville.

The measures taken in by this basin to the north-east are those just described, the rise of the basin to the north, however, carrying the coals higher in the hills and ultimately cutting out the upper ones.

North-west of Brockwayville about two and a half miles, on

Jenkins run, Mr. Powell has opened up a mass of fire-clay and carbonate iron ore balls as follows:

Clay on top, ...	10′	0″
Ferruginous clay, deep red, . . .	5	0
Gray clay slate, ...	3	0
Clay, reddish, with iron ore balls, .	3	6
Clay and clay slate, with some carbonate iron ore balls, ...	20	0
Level of small run.		

The ore balls are so thinly scattered through the mass as to be worthless.

This great mass of clay and clay slate apparently represents the fire-clay deposit so frequently found resting almost directly on top of the Seral conglomerate of XII. It is of course in many cases thinned down and even lacking entirely; but it is a regular and persistent deposit. In this case it seems to hold no fire-clay of sufficient purity to be of any value.

The Little Toby Creek Valley at this point, and to the north-west of it, is filled with great masses of Seral Conglomerate rock, the rounded white quartz pebbles running at times as large as a hen's egg.

On the P. Galusha tract, two and a half miles north-west of Brockwayville, (225 feet above the Little Toby creek,) a coal has been opened up; it is now fallen shut, but apparently did not exceed two feet or two and a half feet in thickness.

Fifteen feet above this coal, in a mass of black slates and shales, there occur six to eight inches of good looking carbonate iron ore in plates, with smaller plates of iron ore above and below, and many ore balls of all sizes scattered through the mass.

A specimen of this P. Galusha iron ore yielded on analysis, (M'Creath):

Iron, - - - - - - - -	27.000
Sulphur, - - - - - - -	.053
Phosphorus, - - - - - -	.108
Insoluble residue, - - - - -	31.120

The ore is hard, compact and silicious, light gray in color, and with conchoidal fracture.

The above analysis shows an iron ore of excellent quality, and it seems to be in sufficient quantity to pay for working, when ready access to market is obtained.

Two coal beds come in above this iron ore. A specimen from the upper one yielded on analysis, (M'Creath):

"Water, - - - - - - -	1.150
Volatile matter, - - - - -	36.000
Fixed carbon, - - - - -	48.099
Sulphur, - - - - - -	7.611
Ash, - - - - - - -	7.140
	100.000

Coke, per cent., 62.85. Color of Ash, red.

The coal is shining, friable, iridescent and showing a large amount of iron pyrites."

The above specimen is merely outcrop coal, but there is not much to hope for from a coal showing such a percentage of sulphur as it does at this point.

West of Brockwayville two and a half miles, on the H. Smith place, a limestone with carbonate iron ore is opened up two feet thick, but the ore is lean. Thirty feet above the limestone, there is a bed of coal two feet thick on the outcrop, with a roof of carbonated clay slate. It is reported a "three foot bed of coal."

Twenty-two feet above this coal is the blossom of a coal bed, not opened up now for measurement, reported as a "four foot coal bed."

The measures at this point are rising to the north-west, to the Fourth anticlinal axis.

At Bovaird's place, three and a half miles west of Brockwayville, Mr. W. H. Frost has opened up a limestone in the run, which shows five to seven feet of hard blue and blue black limestone. It burns well and makes good lime. Directly upon it rests a carbonate iron ore, lean for the most part, weathered brown at the outcrop.

The top layer of this ore is highly fossiliferous, almost the entire mass being made up of closely packed fossil forms.

The fossil ore on top and the ore lying below were analyzed and yielded, (M'Creath):

For the fossiliferous top ore:

"Iron, - - - - - -	36.800
Sulphur, - - - - - -	.034
Phosphorus, - - - - -	.296

Manganese, - - - - - - 1.744
Insoluble residue, - - - - - 22.980

The ore is compact, highly fossiliferous, of a reddish brown color."

The underlying main ore body yielded, (M'Creath):

"Iron, - - - - - - - -	37.700
Sulphur, - - - - - - -	.018
Phosphorus, - - - - - -	.553
Manganese, - - - - - -	2.212
Insoluble residue, - - - - -	20.770"

The ore is compact, fossiliferous and of a reddish brown color.

The above ores are practically the same, and are sufficiently rich in iron to rank as a valuable iron ore.

Forty feet above the limestone a coal bed shows in the ravine, reported two feet thick, with 10 feet of black slate overlying it, holding many iron ore balls.

For convenience for reference tne analyses of coals given above are grouped together in the table below:

	Water.	Volatile matter.	Fixed carbon.	Sulphur.	Ash.	Coke, per cent.
W. M'Cullough	1.32	33.92	53.905	1.505	9.35	64.76
J. Keys	1.36	38.72	53.683	2.047	4.19	59.92
J. J. Stewart	1.83	34.27	58 353	.767	4 78	63.90
W. J. Calhoun	1.20	33.63	55.796	1.504	7.87	65.17
S. Patten	1.87	32.45	61 103	.547	4.03	65.68
P. Galusha	1.15	36.00	48 099	7.611	7.14	62.85

The analyses of the "four foot" coal bed show an excellent coal; rich in gas and with low percentage of sulphur and ash. The detailed description of the openings shows that this bed is the one, and the only one, which is of size and character to ship to market. There is no great difficulty in finding an outlet for it to the Low Grade railroad; but it then comes into competition with the cheaper mining possible upon the Lower Freeport Coal Bed. For the average of this bed is only three and a half to four feet, while the Lower Freeport bed is five and a half to six feet; and the difference in cost of mining would be considerable.

CHAPTER XXVII.

The Fifth or Brookville Coal Basin along the Red Bank Creek.

The Fourth great anticlinal axis crosses the Sandy Lick creek at or near Port Barnet. This axis brings up the measures below the Seral conglomerate, XII, and the country rock along the creek for a few miles is made up of the rocks of X, Upper Catskill, Vespertine. These are usually greenish, yellowish or olive colored sandstone layers, from one to five inches in thickness, finely grained, argillaceous, containing peroxide of iron in hollow concretions. Shales and shaley sandstones, in layers, alternate with the flags and constitute a large proportion of the whole. They separate the formation also into belts, and greatly predominate at its upper limit, where they contain the shelly argillaceous iron ore. From 30 to 40 feet of these upper shales and flags are seen in the valley of the Sandy Lick creek. The above descriptions are taken from Rogers' Final Report of the Geology of Pennsylvania.

The rock exposures at Brookville are not numerous, and the following imperfect section was obtained near the town:

	Hill top.	
	Gray slates, ...	40′
	Coal crop.	
XII.	*Concealed measures*, the surface showing	
XI.	much sandstone, some massive, . . .	193
X.	Thin flaggy sandstone, ...	40
	Level of Red Bank Creek.	

The formation X, Vespertine of Rogers, has been described above; and XI and XII, the Umbral rocks and Seral conglomerate, cannot exceed together 150 feet in thickness, and are probably less than that. Of this 150 feet, less than 50 should apparently be given to the Umbral rocks.

A vertical section made up from Red Bank creek, south of *Fig.* 124. Brookville, shows the measures exposed as follows, (Fig. 124):

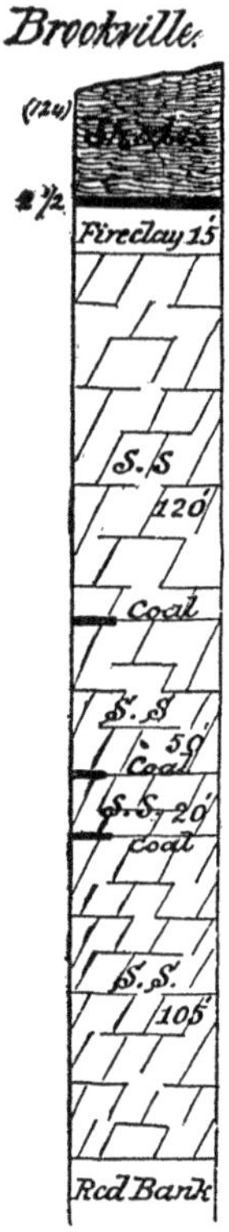

Hill top.	
Shaly slates, ...	25′
Coal, ...	2′ to 3
Fire-clay, ...	15 or more.
Concealed measures, (sandstones?)	120
Small black slate outcrop.	
Sandstone, ...	20
Small black slate outcrop.	
Sandstone and *concealed measures*,	105
Level of Red Bank creek.	

The coal worked on the hill top is hard, bright and tolerably free from iron pyrites. It burns well and answers satisfactorily for domestic use. But it does not rise above a 24 to 30 inch coal, and its use is, therefore, restricted to its immediate neighborhood.

Of the 15 to 18 feet of fire-clay underlying this coal, Wm. Newsome now works the upper six feet. The clay measured about 12 feet when the region was examined, with clay underlying.

The fire-clay is another re-appearance in great thickness and with much purity, of the regular and persistent fire-clay deposit which is found in so many places, widely apart, resting on top of the Seral conglomerate rocks, and is so frequently of sufficient purity for burning.

Three specimens of this Newsome fire-clay were forwarded to the laboratory of the Survey in Harrisburg for analysis and yielded, (S. A. Ford):

	1.	2.	3.
Silica, - - - - -	58.125	60.675	78.075
Alumina, - - - -	26.500	25.915	14.440
Protoxide of iron, - -	3.234	2.210	1.590
Bisulphide of iron, - -	.008	——	——
Lime, - - - - -	.078	.089	.056
Magnesia, - - - -	.555	.465	.480
Alkalies, - - - - -	2.180	1.925	1.670

	1.	2.	3
Sulphuric acid, - -	.058	trace.	trace.
Water and organic matter,	9.725	9.090	4.163
	100.463	100.369	100.474

No. 1 is comparatively soft and of a pearl gray color.

No. 2 is compact, unctuous, and of a pearl gray color.

No. 3 is very sandy, hard, compact, and of a light gray color.

The bricks and ware made from these clays are of good quality and answer well for the purposes to which they are applied in the vicinity. But a comparison of the above analyses, with tables of analyses in chapter XX of this report, shows that they carry a much greater percentage of silica than the Blue Ball, Sandy Ridge and Woodland fire-clays, all of them clays of very high character. The general average of the highest grade Clearfield fire-clays, and the valuable clay from Benezette, in Elk county, (chapter XXI) is about 43 to 45 per cent. of silica and 36 to 39 per cent. of alumina, and about the same for the New Jersey fire-clays given in chapter XX. These Brookville clays possess a much higher percentage of silica and could not be expected to compete with them for the purposes to which those clays are applied. They resemble more closely the English fire-clays, (given in chapter XX).

In the isolated district of coal measures enclosed between the Sandy Lick creek (north fork) and Mill creek, north-east of Brookville, at its north end, a ridge contains the Ferriferous limestone.

In the Fifth Coal Basin, three and a half miles west north-west of Brookville, at Cowan's mine, the coal shows:

Roof, black slate, . . .	3′ or more.	
Coal, bony, . . .	0′ 5″ to 0	6″
Coal, ...	2	3
Slate, ...	0	1
Coal, ...	0	10
Fire-clay floor, . . .	5	0

The coal is hard, bright and good, and is dipping gently to the north-west. It is apparently the Clarion coal.

At Kennedy's mine, one-half mile north of Cowan's mine, the coal shows:

Roof, black slate		
Coal, ...	1′	4″
Slate, ...	0	1

Coal, ...	0′ 10″
Fire-clay floor.	

The coal dips gently to the north-west.

At Morrison's mine, three-fourths of a mile west of Kennedy's mine, the coal shows:

Roof, black slate.	
Bone coal, ...	0′ 3″
Coal, somewhat bony, . . .	0 6
Coal, ...	2
Slate parting, ...	0 1
Coal, ...	1 0
Fire-clay floor, ...	4 or more.

This coal is at a lower level than Cowan's mine, (it lies down the dip of the measures,) but resembles it closely and is apparently the same bed.

A small 2 foot 6 inch thick coal bed is opened up 20 feet above the Morrison mine. It has a black slate roof and a fire-clay floor.

At Brown's mine, one-half mile east of Kennedy's, three miles north-west of Brookville, the coal shows:

Roof, black slate.	
Coal, ...	2′ 6″ to 3′
Fire-clay floor.	

Truby's old coal mine, just north-west of Brookville, and 285 feet by barometer above Red Bank creek, showed only "15 to 18 inches" of coal. It is now fallen shut.

Fig. 125.
1 mile below Troy.

It will be seen from the above description of the mines now working, that the Brookville and Clarion coal beds through this region do not average more than from 30 to 36 inches in thickness; and though the coal is of excellent character, it is evident that these beds can never compete with better situated and larger coal beds for an open market. Their use is restricted to the local demand, in this settled country a very considerable quantity; and this local demand they are entirely fitted to supply, as they cover a reasonably large area and are of good quality.

Going down the Red Bank creek, at a point two and a half miles east of Troy, the ferriferous limestone appears, thin and blue, with a sandstone roof. Twenty feet below it is the Clarion coal bed, one and a half feet thick. It occurs again in the little run ascending the hill to Troy. The Clarion coal, opened below it in several places, is two feet thick. The Kittanning coal was not discovered above the limestone. (Lesley in F. R., 1858.) One mile below Troy the following vertical section (Fig. 125) was obtained by Lesley, (Rogers' Final Report, 1858):

Hill top.		
Sandstone, Freeport, ...	50′	0″
Shale and sandstone, ...	75	0
Coal, Kittanning, ...	2	0
Brown shale with compact oolitic iron ore, ...	30	0
Ferriferous limestone, ...	4	0
Shale and slate, (ore in lower part,)	30	0
Coal, Clarion, ...	1	2
Shale, ...	55	0
Coal, Brookville, ...	3	6
Shale and slate, ...	45	0
Level of Red Bank creek.		

The upper part of the section in Fig. 125 is compiled.

The Lower Freeport Coal Bed is first met with capping the highest hills on the Red Bank creek, six miles below Troy. It is opened four or five miles below, nearer Kittanning, as in the following vertical section (Fig. 126):

Fig. 126.

Hill top.		
Shales, ...	40′	0″
Coal, Upper Freeport (Middle?)		
Limestone, Freeport.		
Sandstone, ...	50	0
Coal, Lower Freeport, ...	4	6
Sandstone, Freeport, a few pebbles,	50 to 60	0
Shales, etc., ...	75	0
Coal, Kittanning, ...	2	6
Sandstone, flaggy, ...	45′ to 50	0
Dark shales, ...	10	0
Ferriferous limestone, ...	6	0
Shales, olive and yellow, . . .	20	0
Coal, Clarion.		
Level of Red Band creek.		

(Rogers' Final Report.)

The geographical position of all these vertical sections is shown on the map (Plate VIII) accompanying this report.

The first extensive opening made on the Lower Freeport Coal Bed, and from which coal is shipped to market, is at the Fairmount colliery, at Fairmount station, on the Low Grade railroad, one mile east of New Bethlehem. The Lower Freeport coal is here in the Fifth basin, as in the Fourth basin at Reynoldsville, a large, fine bed, superior to the other coal beds in size and character, and therefore the only one now worked to any extent. The beds above and below are worked in a small way in a few places.

A vertical section was made (with the barometer) at Fairmount, the outcrops of the different coals showing either as benches or in trial pits; and a second vertical section was also made just east of Fairmount station, on the opposite side of Middle Run. From the exposures of these two sections the vertical section (Fig. 127) is compiled:

Fig. 127
Fairmount.

Hill top.		
Gray slates,	30′	0′
Coal,	4	0
Concealed measures,	60	0
Coal, Lower Freeport,	6′ to 7	0
Sandstone, Freeport,	47	0
Coal, Kittanning,	4	0
Concealed measures,	31	0
Ferriferous limestone and Clarion iron ore on top,	5	0
Thin sandstones and *concealed measures,*	64	0
Coal,	2	6
Measures, partially seen, some sandstone,	33	0
Coal,	4	0
Concealed measures,	6	0
Limestone,	4	0
Concealed measures,	10	0
Level of Red Bank creek.		

The face of the hill is covered with loose stuff, preventing an accurate determination at this point of the character of the interval rocks.

The section of the measures above water level is supplemented by an imperfect record of an oil well started from just above the level of Red Bank creek, about one mile east of Fairmount station. The record shows that a coal was struck at 142 feet below the surface; the rocks not recorded, but "mainly sandstones."*

*This well has since been abandoned. It is reported to have reached a depth of 1,450 feet and "no third sand."

The lower limestone of the section (Fig. 127) at Fairmount, is not opened up for working, but has been sufficiently exposed in the past for measurement, giving about four feet of limestone on the average.

The coal over this limestone, at the grade of the railroad, is not opened for working at this place. At one time it was worked to a small extent a short distance east of Fairmount station. The coal shows:

Black slate overlying with many balls of carbonate of iron,	25′	0′
Roof, black slate,	3	6
Coal.	3′ 6″ to 4	0
Slate parting, pyritous,	0	4
Coal,	0	6
Fire-clay,	3	0
Massive whitish sandstone,	6′ to 7	0

The coal shows much iron pyrites through it.

A specimen of this coal (Brookville coal) yielded, on analysis, (M'Creath):

"Water,	1.370
Volatile matter,	37.680
Fixed carbon,	39.353
Sulphur,	8.427
Ash,	13.170
	100.000

Coke, per cent., 60.95. Color of Ash, rea.

The coal has a dull lustre outside, resinous inside, with pyrites and a large quantity of sulphate of iron."

This specimen was necessarily taken from near the outcrop, and the percentages of sulphur and ash are undoubtedly above the average; but the coal is clearly both sulphurous and ashy.

The Clarion coal is only picked into on the outcrop at Fairmount, sufficiently to show the position of the bed and its general thickness (two and a half feet.)

The Ferriferous limestone and ore are not worked at Fairmount, but are worked at Himes' ore bank at New Bethlehem, and will be described as measured there. The Kittanning coal is also worked at Himes' place and could be measured and examined there.

Fig. 128.
Fairmount

The Lower Freeport coal bed is opened and worked at Fairmount colliery, on the north bank of the Red Bank creek. The coal as measured in the mine reads (Fig. 128) as follows:

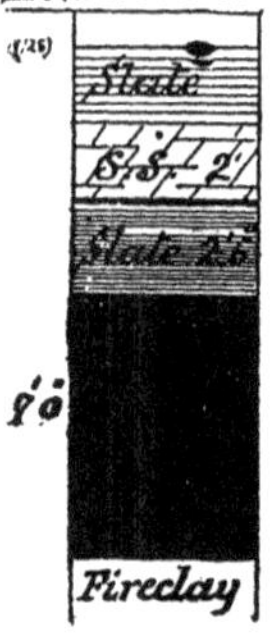

Sandstone, ...	2′ 0′
Roof, black slate and cannel slate, . .	2 6
Coal, ...	5′ 6″ to 7 0
Fire-clay floor, ...	1 6

Though the normal roof of this fine bed of coal is a black slate, yet for 100 to 200 yards in from the outcrop there were found about two and a half feet of fire-clay resting directly on top of the coal, and the black slate above that. While this abnormal roof continued the coal was less regular in size, the clay occasionally swelling down and cutting out one to two feet of the coal. The floor is usually very smooth and regular, only two or three small swamps having been encountered in the mine, in no case, however, throwing the level out more than three or four feet. The mine runs in north-west and drains, showing a gentle dip to the south-east. 18 feet of gray slates overlie the roof of the coal at the mine mouth to the hill top just above.

Two specimens of this Fairmount Mine Lower Freeport Coal yielded, on analysis, (M'Creath):

	1.	2.
"Water, - - - - -	1.700	1.320
Volatile matter, - - -	38.930	40.800
Fixed carbon - - -	56.096	52.879
Sulphur, - - -	.604	.881
Ash, - - - - - -	2.670	4.120
	100.000	100.000

No. 1. Coke, per cent., 59.37. Color of Ash, cream.

The coal is bright, resinous, hard, compact, with a small amount of pyrites.

No. 2. Coke, per cent., 58.88.

The coal has a dull resinous lustre, containing veins of slate and considerable iron pyrites. Forms a coherent coke with metallic lustre. Gray Ash, with reddish tinge."

The above analyses shows a most excellent coal; with a very high percentage of volatile matter, equal in gas percentage to the Westmoreland County or Youghiogheny gas coals from the great Pittsburg Bed, and with a very low percentage of sulphur and ash.

This high character of the coal, together with the fact that the whole thickness of the bed, seven feet, is not parted by any regular and persistent slate partings, make the coal deposit one of exceptional value. The character of the coal will not probably run, on the average, quite up to the high standard given by the analysis, but the mine always yields a good coal.

A contoured map of the Fairmount Coal Company's property is appended to this report (Plate IX); the outcrop line of the Lower Freeport Coal Bed being laid on plainly, following around the hill sides. It serves to show at a glance how the coal lies in the hills in this Fifth Basin, and to illustrate the broad general proportion of coal washed away from valleys and existing as outcrop coal; illustrating how the general estimates of quantity of coal gone, as compared with the total quantity indicated by square miles of coal area, have been got in our statements respecting the Clearfield and Reynoldsville Coal Basins.

The colliery ships 200 to 250 tons a day, partly for gas coal to Buffalo and the north-west, or *via* Red Bank and the Allegheny Valley Railroad to the Oil Regions, the coal bearing a deservedly good name for gas producing power. And also ships partly for steam raising purposes; the coal,burning as a strong steam coal, and working well in the locomotive engines of the Low Grade railroad.

It is noteworthy, in view of the cannel nature of this Lower Freeport coal bed, south of the Red Bank creek, that occasionally considerable quantities of this Fairmount coal take on a dull, semi-cannel appearance, this change more usually occurring in the upper three feet of the bed. This perhaps accounts for the high percentage of volatile matter. The cannel slate roof shows no fossil plants.

The coal bed above the Lower Freeport coal, known usually as the Upper Freeport bed through the region, is not opened on the Fairmount property; but it has been opened and worked in the past at several places on the hill tops on the east side of Middle run, east of Fairmount station. It yielded, according to description, (for the mines are now fallen shut) a good average four feet of coal, and of decidedly good quality.

In comparing the sections made south of Red Bank creek, it will be shown that the bed is more probably the Middle Freeport coal bed.

Where the ferriferous limestone and carbonate iron ore have been opened up on Middle run, about one and a half miles north of Fairmount station, the limestone measured four and a half to five feet of good blue limestone. The carbonate iron ore (Clarion ore), resting directly upon it, ranged from six or eight inches to twelve inches or more in thickness. Clay slates, for eight or ten feet, overlie the ore, with ore balls through them.

At New Bethlehem a vertical section (by barometer) from the Red Bank creek to the hill top on the north side of the Red Bank reads (Fig. 129) as follows:

Fig. 129.

Hill top.	
Gray slates, ...	40′
Black slate outcrop.	
Thin sandstones and slates, . .	62
Bench.	
Sandy slates, ...	70
Bench, (Lower Freeport)	
Concealed measures, ...	65
Coal, Kittanning, ...	4
Clay slates, ...	40
Ferriferous limestone and Clarion iron ore on top, ...	5
Concealed measures, ...	55
Coal smut?	
Concealed measures, ...	50
Level of Red Bank creek.	

The coal below the Ferriferous limestone is not opened and is very slightly exposed, the gentle sloping hill side being covered deep with loose surface stuff, preventing for this lower part of the section any accurate determination of the character of the rocks.

Mr. Himes has opened up the Ferriferous limestone and iron ore just north of the village of New Bethlehem.

The limestone shows six feet of thick bedded, dove colored and bluish limestone, of fair quality.

The iron ore rests directly on top of it and shows one foot thick. It varies in thickness much and quickly, running above and below the above thickness, but apparently averaging about that amount of ore.

This iron ore is well and widely known through the region west of this place, and is extensively mined at numerous points.

Over the ore are four feet of partially decomposed clay slates and fire-clay, with many carbonate iron ore balls of all sizes. These ore balls vary very much in quantity in different parts of the same mine; but as the clay is naturally the easiest rock to work, in driving the gangways the solid ore is near the bottom of the drift, and the driving is done through the overlying clay, bringing out with the clay such ore balls as may be found. It makes a variable but considerable addition to the output of the mine.

A specimen of this Clarion iron ore from the Himes mine forwarded to the laboratory of the Survey in Harrisburg, yielded on analysis (M'Creath):

"Iron, - - - - - - -	30.600
Sulphur, - - - - - - -	.075
Phosphorus, - - - - - -	.225
Insoluble residue, - - - - -	18.300

The ore is hard, compact, of a dark gray color and conchoidal fracture."

The above analysis corresponds well with the known character of this deservedly esteemed ore. It shows a good percentage of iron and very little sulphur and phosphorus; and is, when roasted, an ore rich in metallic iron.

The Kittanning Coal Bed is opened and worked by Mr. Himes 45 feet above the iron ore. It shows:

Dark slate, with some ore nodules, . .	12′	0″
Roof, black clay slate, ...	1	0
Bony coal, ...	0	5
Coal, hard, ...	2	2
Fire-clay floor.		

The coal, though rusty where examined, is hard and good, and answers very well for domestic use in the village and vicinity. The coal is rising gently to the north-west.

None of the benches above the Kittanning coal are opened on this hillside.

While the measures between the Red Bank Creek and the Ferriferous limestone and its iron ore do not appear in the section (Fig. 129) at New Bethlehem, yet some parts of them show on the railroad just west of the village. The railroad here cuts through the top of the Clarion sandstone. As the railroad is sinking in level slightly to the westward down the stream, while the measures are rising gently to the north-west to the Fifth anticlinal axis, the sandstone is seen rising steadily above the railroad and mounting higher and higher in the hills. Over the sandstone at New Bethlehem is a fire-clay and a small coal smut. The sandstone shows much current bedding, and is a massive whitish and grayish rock, 40 to 50 feet thick.

Below this Clarion sandstone and between it and the massive sandstone of the Seral conglomerate there come in softer grayish slates and thinner sandstones, 100 or more feet of them in all. Only one small coal smut shows in them, and that well down towards the Seral conglomerate.

Going north from New Bethlehem, up Leasure Run, a coal is opened at Lowries' mine, two-thirds of a mile north of the village. It shows:

Roof, black slate.	
Coal, ...	4′
Fire-clay floor.	

The coal is only mined into near the outcrop, but it appears hard and good.

The vertical section as seen at this point reads:

Hill top.	
Concealed measures, ...	50′
Small bench.	
Concealed measures, ...	52
Small bench.	
Concealed measures, ...	48
Coal, Lowrie's, ...	4
Concealed measures, massive sandstone on surface, . .	83
Bench, coal.	
Concealed measures, ...	120
Bench.	
Concealed measures, ...	50
Level of Leasure creek.	

The hillside is covered deep with loose stuff, especially sandstone pieces brought down from the sandstone below Lowries' coal, and the character of the interval rocks is obscured.

At Himes' place, 2 miles north of New Bethlehem, a coal has been opened but is not now worked. Below it are several feet of fire-clay which has been used for making crockery ware, and answered well.

At Space's place, $3\frac{1}{2}$ miles north of New Bethlehem, tne Ferriferous limestone has been opened up, 4 to 6 feet thick, with its regular carbonate iron ore on top, keeping about the average thickness.

The Thomas Ditty Mine is opened and worked on the Lower Freeport Coal Bed about four miles north of New Bethlehem. The coal as measured in the mine shows (Fig. 130) as follows:

Fig. 130.

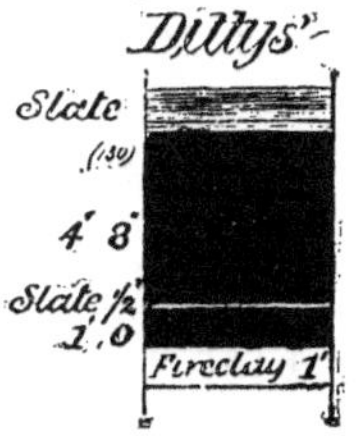

Roof, black slate, smooth.		
Coal,	4′	8″
Slate parting,	0	$0\frac{1}{2}$
Coal,	1	0
Fire-clay floor.		

A light colored sandy clay underlies the mine floor. The coal is only mined to a small extent, and the mine is not yet driven in for anv distance. But the coal looks good and bright.

A vertical section at this point reads:

Hill top.		
Bench.		
Concealed measures,	70′	0′
Coal, (L. F.) Ditty's,	5	6
Sandstone,	63	0
Coal, not opened, (Kitt.)		
Concealed measures,	75	0
Bench.		
Concealed measures,	70	0
Bottom of hill.		

The same Lower Freeport Coal Bed is opened at M'Nulty's mine, four miles north north-west from New Bethlehem. The coal as measured in the mine shows (Fig. 131) as follows:

Fig. 131.

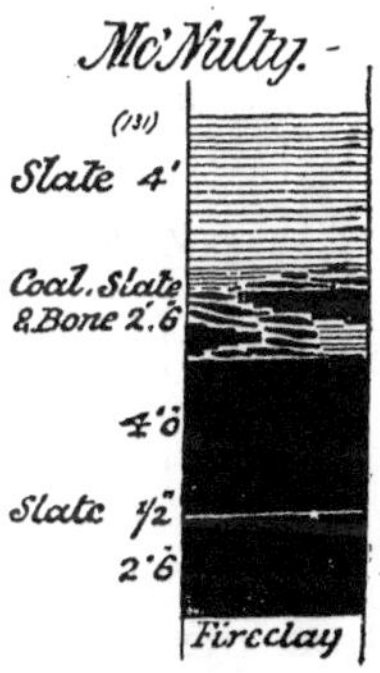

Roof, thin slates,	4′	0′
Coal, with bone coal and slate,	2	6
Coal,	4	0
Slate parting, persistent,	0	$0\frac{1}{2}$
Coal,	2	6
Fire-clay floor.		

The mine is not driven in far; but the coal looks good and bright so far as worked.

The measures are rising to the north-west. The Freeport sandstone, massive and somewhat conglomeritic in places, covers the hillside in blocks below the mine.

The bench of the Upper coal is marked, and about 70 to 75 feet above the M'Nulty mine.

At Hamm's store, seven miles north north-west of New Bethlehem, a limestone has been opened (the lower limestone of the section in Fig. 127). The Lower Freeport Coal Bed is in the hill tops east of this place, but it is high up on the hills and from thence passes out to daylight rising up north-west to the Fifth anticlinal axis. There is apparently no hill crest which takes in the Freeport Bed between this hill east of Hamm's store and the Fifth axis.

The Wilkins mine is opened on the Lower Freeport Coal Bed three miles north north-west of New Bethlehem. The coal as measured in the mine shows (Fig. 132) as follows:

Fig. 132.

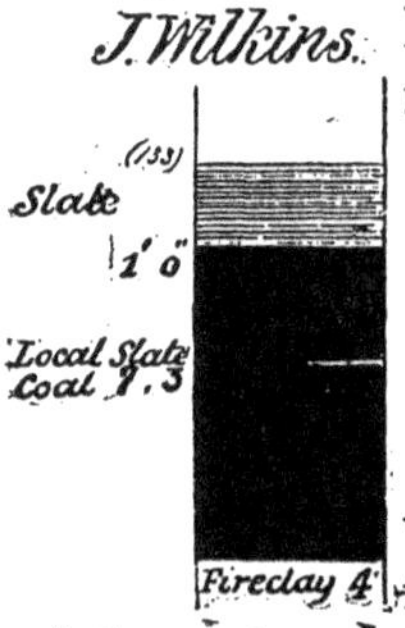

J. Wilkins.

Roof, black slate.		
Coal, with some bone coal,	1'	0"
Coal, ...	2	0
Slate parting, . . .	0	0½
Coal, ...	5	3
Fire-clay floor, hard, . .	4.	0 or more.

This is an extremely handsome exhibition, with a firm dry roof and even and regular floor. The coal mines out well, hard, firm, bright and very free from iron pyrites. The coal shows the same dull look, inclining to a cannel structure, which has been described as characterizing the Fairmount mine, which it resembles closely in size of bed and appearance of coal. The measures are rising gently to the north-west.

Three miles north north-west of New Bethlehem, the two Freeport Coal Beds were opened and worked for St. Charles' furnace.

The mine on the Lower Freeport Coal Bed is now fallen entirely shut and cannot be measured.

The opening on the bed above is 60 feet higher on the hill-side and measured:

Roof, black slate.	
Coal, ...	3' 9
Fire-clay floor.	

The coal comes out well, looking good and free from pyrites.

It is reported that the coal from this upper bed was used raw at St. Charles furnace; while the coal from the Lower Freeport bed required to be coked.

Sandstone lumps and boulders, a fine grained conglomeritic rock in places, extends from the bench of the Upper Bed to the hill top, 50 feet above.

Shape of the Fifth Basin.

The Fifth Coal Basin is very free from any geological complications.

Its eastern side is marked by the Fourth anticlinal axis which crosses the Sandy Lick creek at Port Barnet, two miles east of Brookville; and its western side is bounded by the Fifth anticlinal axis, which crosses the Red Bank creek at Lawsonham.

The measures slope gently down into the basin from these anticlinal axes; the centre of the basin lying not far from the Fairmount colliery.

The Lower Freeport coal occupies the hills in this centre of the basin. But it is rising steadily, with the measures, to the north; and therefore shoots out of the hill not far beyond the points described in this report.

Sinking to the south-ward, with the measures, these coals pass through the high land between the Red Bank and Mahoning creeks, and pass on in the hills south of the latter stream.

New Bethlehem and Red Bank Coal.

A vertical section of the measures showing on the south side of Red Bank creek, opposite New Bethlehem, in Armstrong county, leveled by barometer, reads as follows (Fig. 133)

Hill top.	
Concealed measures, ...	50
Bench.	
Sandstone, massive, . . .	55
Coal, ...	3 or more.
Sandy slates, ...	65
Coal, (L. F.,) not opened.	
Sandstone, massive, (Freeport,) .	40
Steep hill side, with nothing showing,	200
Level of Red Bank creek.	

The Big Bed of coal (Lower Freeport) has been opened up along the hill face, but the coal could not be examined and measured in any of the old openings. In one place it was only "3 feet thick;" in another place,

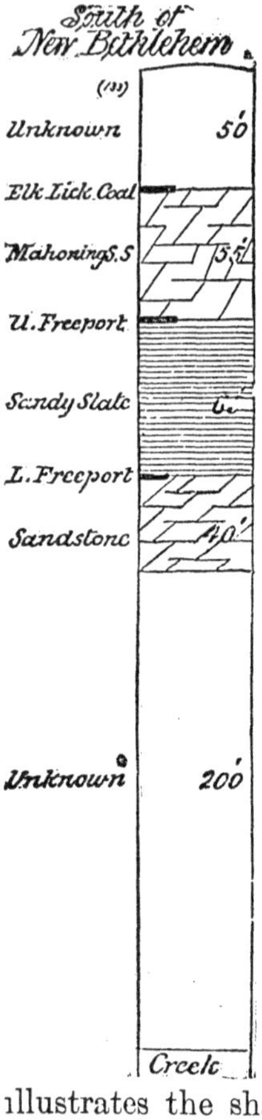

Coal,	2′	
Clay parting,	2	6
Coal,	2	6

and in a third opening, driven in 400 feet, the coal is reported as "6 foot."

Fig. 133.

The coal bed 60 to 65 feet above has been opened up on its outcrop and showed:

Roof, black slate.	
Coal,	3′
Fire-clay floor, soft,	2 or more.

The sandstone boulders which cover the hill side, above the bench of this upper bed, are in places decidedly conglomeritic.

The Red Bank Coal and Mining company have opened and are working the Freeport coals at their colliery, two miles south-west of New Bethlehem. The Red Bank creek makes the boundary line between Clarion and Armstrong counties and the colliery therefore lies in the latter county.

The contoured map of the Red Bank Coal company's property appended to this report (Plate X) is furnished by John F. Blandy, mining engineer, the present superintendent of the company. The outcrop line of the upper Coal Bed is laid on the map plainly, following around the hill sides, and the map illustrates the shape of the country, loss by erosion, &c. A cross section accompanies the map, showing the thickness of the measures above water level and the geological structure.

The vertical section of the measures developed is also furnished by Mr. Blandy, (Fig. 134,) which will be found drawn out in full on Plate X, appended to this volume. It is drawn in detail and the heights were taken with a spirit level.

The ferriferous limestone and the ore on top are not worked on this property, but the ore has been, keeping a full average thickness about one foot.

The Lower Freeport coal is opened and extensively worked by the company.

Where measured in the mine the coal shows, (Fig. 135) as follows:

Fig. 135.

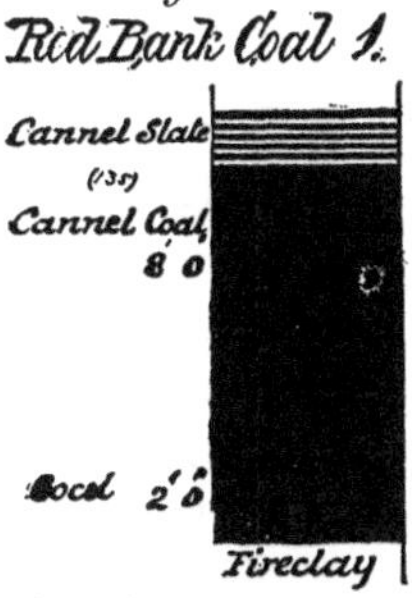

Red Bank Coal 1.

Cannel Slate (1'3?) Cannel Coal 8 0 Coal 2' 0 Fireclay

Roof, cannel slate	
Cannel coal, ...	8
Bituminous coal, . . .	2
Fire-clay floor.	

The cannel coal keeps its thickness along the main entry, but in driving off cross headings it is found to thin down on both sides, finally running out to nothing.

Two specimens of this cannel coal from the Red Bank mines yielded on analysis, (M'Creath):

	No. 1.	No. 2.
"Water, - - - - -	0.510	.730
Volatile matter, - - -	30.490	31.680
Fixed carbon, - - -	46.194	49.815
Sulphur, - - - - -	.576	.455
Ash, - - - -	22.230	17.320
	100.000	100.000

Specimen No. 1, has a conchoidal fracture, is very hard, with considerable carbonate of lime through it.

No. 2, has dull resinous lustre, is hard and compact, with conchoidal fracture, and contains considerable carbonate of lime.

In coking the particles do not seem to fuse together very thoroughly, and the resulting coke is only slightly coherent. The Ash is yellowish white."

A specimen of the bottom bench of this coal bed, the two foot bituminous coal, yielded on analysis (M'Creath):

"Water, - - - - - - -	1.650
Volatile matter, - - - - - -	39.120
Fixed carbon, - - - - - -	52.716
Sulphur, - - - - - - -	2.634
Ash - - - - - - - -	3.880
	100.000

Coke, per cent., 59.23. Color of Ash, brown.

The coal is bright, has a resinous lustre, with considerable iron pyrites in veins."

The above analyses show that the bituminous coal in the floor carries more volatile matter than the cannel coal, but has also a large percentage of sulphur.

Thin shales and slates overlie this cannel coal bed; and 25 feet above it the company have opened and worked another coal bed which reads (Fig. 136) as follows:

Fig. 136.

Roof, cannel slate.		
Coal, ...	0′	4″
Slate, ...	0	1
Coal, ...	3	7
Floor, in one case clay, in another sandstone.		

The bed is rolling and irregular.

A specimen of this coal yielded on analysis (M'Creath):

"Water, - - - - - - -	1.690
Volatile matter, - - - - - -	35.940
Fixed carbon, - - - - - -	53.950
Sulphur, - - - - - - -	3.380
Ash, - - - - - - - -	5.040
	100.000

Coke, per cent., 62.37. Color of Ash, gray.

The coal is bright, shining, friable, containing considerable pyrites and charcoal."

The analysis shows a good gas yielding coal, but very sulphurous.

Forty-five feet above this coal, and 69 feet above the Lower Freeport (cannel) coal bed, the company have opened and are working a coal bed. The coal shows where measured in the mine (Fig. 137) as follows:

Fig. 137.

Red Bank Coal 3

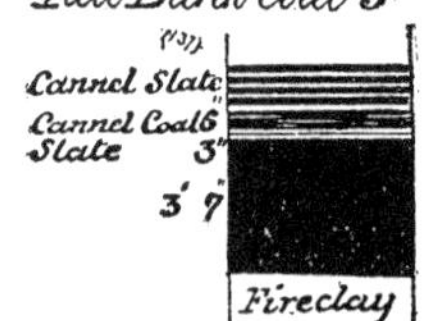

Roof, cannel slate.		
Coal, bony, ...	0′	6″
Slate, ...	0	3½
Coal, ...	3	8
Fire-clay floor, hard, .	3	0 or more.

The mine is much troubled by fire-clay horsebacks and has an uneven and rolling floor.

This upper coal is called the Red Bank "Orrel" coal; and a specimen of it yielded on analysis (M'Creath):

"Water, - - - - - - -	1.840
Volatile matter, - - - - - -	35.940

Fixed carbon, - - - - -	53.661
Sulphur, - - - - - -	1.739
Ash, - - - - - -	6.820
	100.000

Coke, per cent., 62.22. Color of Ash, gray.

The coal is bright, very hard, with considerable iron pyrites and mineral charcoal."

The cannel coal (Lower Freeport) is opened at the Widow Thompson mine, about three and a half to four miles east of Red Bank colliery. The coal as measured in the mine shows (Fig. 138) as follows:

Fig. 138.
THOMPSON MINE.

Coal, bituminous, in roof .	1′	0″
Cannel coal, ...	7	0
Slate, ...	0	6
Coal, bituminous, in floor.		

The cannel coal seems at times to almost run into a bituminous coal for layers of six or twelve inches, keeping, however, a cannel character, though not specially rich in volatile matter.

At the end of the main heading a small pinch was beginning, the coal being somewhat crushed and materially cut down in size. This was not cut through; it is probably only a small roll.

A specimen of this Thompson coal, forwarded to the laboratory of the Survey in Harrisburg, yielded on analysis (M'Creath):

"Water at 225°, - - - - - -	1.540
Volatile matter, - - - - -	36.730
Fixed carbon, - - - - -	53.210
Sulphur, - - - - - -	.630
Ash, - - - - - -	7.890
	100.000

Coke, per cent., 61.730.

The coal has a resinous lustre, is hard and compact, with no visible pyrites. In coking it forms a coherent but somewhat porous coke of a dull lustre, and yields a yellowish white ash."

At the Thompson mine a coal has been opened 102 feet above

the cannel (Lower Freeport) coal. A bench occurs between the two, but is not opened.

This corresponds with the average section of the Freeport beds. In the measures below the Freeport bed there is much variation in thickness; the Seral Conglomerate running down in thickness from 250 feet on the Allegheny mountain escarpment, and the same thickness at Karthaus in the Western division of the First Coal Basin, to 100 feet or less in thickness in the Fourth and Fifth Coal Basins. But the distance from the bottom of the Seral Conglomerate rock to the Lower Freeport bed is scarcely ever much changed; the upper part of the conglomerate being represented by thin sandstones and gray slates, holding coal beds, small usually, but sometimes valuable.

With the Freeport coals, however, there comes in an era of stability in the measures; and over long stretches of country the Freeport beds appear, thickening and thinning very much in thickness of coal, but usually regular as to rock partings.

The three Freeport beds commonly include about 100 feet of measures; and for this reason the bed 102 feet above the Thompson mine is apparently the Upper Freeport Bed.

This would make the upper bed at Red Bank colliery and at the Fairmount colliery, the Middle Freeport coal; and the bed 25 feet above the Lower Freeport coal at the Red Bank mines, in one case the roof of the the lower coal being only 13 feet below the uneasy floor of the coal above, a sporadic and nameless deposit.

The cannel coal itself is clearly a pot deposit. The workings and outcrops indicate that it runs for several miles as a narrow trough of cannel coal, thinning away at the sides. This only applies to the cannel coal deposit which makes the upper bench; the lower bench, two feet thick, of bituminous coal is found regular and smooth, outcropping as a coal bed, without any cannel. Mr. Blandy shows this structure of the cannel bed on his cross section. This would reduce the Lower Freeport Coal Bed, where the cannel bench is gone to only a two-foot bed, and leans to the conclusion that these measures just here are abnormal, and that this cannel coal bed and the bed just above represent together the Lower Freeport Coal Bed. It must be noted that where opened at the Thompson mine, the cannel coal shows much less percentage of ash; and that it varies much in a single mine, the specimens from Red Bank colliery, giving in one case 22 per cent., in another 17 per cent. of ash.*

The coal beds of the section given above, pass through the hill which separates the Red Bank and Mahoning creek, a distance of two and a half to three miles at this point, and come

*Mr. Binney, F. R. S., has been for many years investigating the structure of coal, and he thinks that he has established the following facts: Soft caking or cherry coal, is chiefly composed of the bark, cellular tissue, and vascular cylinders of coal plants, with some macrospores and microspores.

Coking coal has much the same composition, except that it contains a greater proportion of bark.

Splint or hard coal, has a nearly similar composition, but with a great excess of macrospores [large seeds.]

Cannel coal, especially that yielding a brown streak, is formed by the remains of different portions of plants which had been long macerated in water; it contains a great excess of microspores [small seeds.]

(Mr. Leo Lesquereux, the eminent fossil botanist, in his Slippery Rock report, gives the following description of the constituents of ordinary bituminous and cannel coals:

"When the combustible matter has been formed especially from the remains of aerial plants, whose tissue was mostly vascular, or vascular and cellular, like that of the *Lepidodendron*, *Sigillaria*, ferns, &c., it becomes by mineralization a hard coal, with thin layers or distinct laminas, sometimes shining, sometimes mixed with opaque layers and flakes of charcoal, and giving by combustion, a proportion of ashes, according to the nature of the wood. When it has been formed merely by floating fresh water vegetables, like *Stigmaria* and its leaves, the compound, originally half fluid, and more easily decomposed, becomes, by the slow process of combustion, compact, homogeneous, without apparent layers, tending to mere bitumen, thus forming the different varieties of cannel coal.")

Macrospores are from one-twentieth to one-twenty-fifth of an inch in diameter, and can easily be seen by the naked eye. Their exterior is composed of a brown bark-like substance, containing within it carbonate of lime, or bisulphide of iron, according to the nature of the matrix. The microspores are about three hundred and twenty times less in size, and contain some form of hydrocarbon, which, by the action of heat, becomes paraffin. These conclusions were arrived at merely as to the composition of the different kinds of coal. Each seam is materially affected by the nature of the roof, since, if it is an open sandstone, gaseous matter can freely escape, which is, of course, not the case when the seam is roofed in with air-tight black slate or blue bind.

The above digest of Mr. Binney's conclusions is taken from the New York *Engineering and Mining Journal*, Mr. Binney's work not being accessible at the time of writing.—F. P.

out to daylight on the north bank of the Mahoning, but are not opened up there for working.

Mr. Hamilton has opened one coal bed on the outcrop, which shows:

Sandy fire-clay, overlying, ...	10′	0″
Carbonated clay, ...	1	0
Coal, ...	3	10
Floor not seen, reported fire-clay.		

It is reported that limestone was once opened just below this fire-clay floor.

If this be the Lower Freeport coal bed, it has gone down in thickness from seven feet to less than four feet; or it may be the Middle Freeport coal bed, with its limestone in place. A coal bench shows on the hill side about 70 feet above.

The detailed description shows that this Fifth Coal Basin along the Red Bank creek possesses great natural advantages, and must soon be developed, now that the opening of the Low Grade railroad has given it access to market.

The Clarion iron ore is regular in size and character, and lying low in the hills in the centre of the basin has lost little by erosion.

The Lower Freeport is a noble bed, of unusual size and purity; and, to judge from the roughly made coke, done in the open air without ovens, at Fairmount colliery, is evidently capable of furnishing a coke of high quality for iron making.

Limestone is abundant, with the iron ore.

The Freeport Coal Beds occupy the hills in the centre of the basin, covering a region several miles wide, this area growing narrower to the north-east as the measures rise in that direction. The Freeport beds do not extend more than eight miles north-east of the Red Bank creek; but go through the hills to the Mahoning creek and on to the south-west of that stream.

Descending Red Bank creek the Fifth anticlinal axis crosses the stream at about Lawsonham. The Freeport sandstone is the surface rock along a strip of country where the Fifth axis runs south-west after crossing the Red Bank creek. The following vertical section (from Rogers' Final Report) was made at the mouth of Mahoning creek:

Fig. 139.

Mouth of Mahoning Creek.

Ferriferous limestone,		15′	0″
Shales, with ore balls,		35	0
Coal, Clarion,		2	6
Shales, etc.,		20	0
Coal, Brookville,		1	0
Sandstone, massive,		60	0
Shale, silicious,		25	0
Olive bituminous shales,		15	0
Coal,		1	6
Seral Conglomerate, massive, also shaly, XII,		100	0
Shale, sandy, partly carbonaceous, with seams of calcareous sandstone from one-quarter inch to one inch thick,	XI.	20	0
Bituminous shale,	XI.	0	3
Coal, Sharon,	XI.	0	2½
Shale, sandy above, bituminous below,	XI.	3	6
Coal,	XI.	0	6
Thin bituminous slate, with some silicious layers,	XI.	11	0
Coal,	XI.	0	1½
Blue, sandy clay,	XI.	2	0
Slaty sandstone,		25′ to 30	0
Level of Mahoning creek, X.			

(Rogers' Final Report.)

The Freeport limestone and the Upper (Middle?) Freeport Coal Bed are seen on Scrubgrass creek, which enters the Mahoning creek two miles above its mouth.

For convenience of reference the analyses of coals in the Fifth Coal Basin are given below in tabular form:

	Water.	Volatile Matter.	Fixed Carbon.	Sulphur.	Ash.	Coke.
Fairmount, Big Bed	1.320	40.800	52.879	.881	4.120	58.880
Fairmount, Big Bed	1.700	38.930	56.096	.604	2.670	59.370
Fairmount, Lower Bed	1.370	37.680	39.353	8.427	13.170	60.950
Red Bank Colliery, Cannel	.510	30.490	46.194	.576	22.230	69.000
Red Bank Colliery, Cannel	.730	31.680	49.815	.455	17.320	....
Red Bank Colliery, below Cannel	1.650	39.120	52.716	2.634	3.880	59.230
Red Bank Colliery, Middle Bed	1.690	35.940	53.950	3.380	5.040	62.370
Red Bank Colliery, Upper Bed	1.840	35.940	53.661	1.739	6.820	62.220
Thompson Mine	1.540	36.730	53.210	.630	7.890	....

The prevailing gas coal character of the above analyses is a marked characteristic; and in this basin, as in the Reynoldsville Gas Coal Basin, this special character is found running through all the coal beds of the measures exposed, and not confined to one bed alone.

APPENDIX.

Extracts from H. D. Rogers' Memoir on the Coal Formation of Pennsylvania, on pages 796–810, *of the Final Report of the Geology of Pennsylvania, published in* 1858.

Extent of the Coal Area.

Considering all of the outlying portions of the formation as subordinate and intimately connected parts of one great bituminous coal-field, the S. E. boundary of which is the escarpment of the Allegheny and Cumberland mountains, the dimensions of the great basin will be nearly as follows: its length, from N. E. to S. W. is rather more than 720 miles, and its greatest breadth about 180 miles. Upon a moderate estimate, its superficial area amounts to 70,000 square miles.

Though the deep anthracite basins abound in curious structural features, and contain thick seams of coal, they chiefly interest us at present by the geographical position which they occupy. More than 40 miles distant from the general denuded margin of the main or W. coal-field, they nevertheless present, in the character of their strata, and of the rocks upon which they repose, unequivocal evidence that they and the bituminous basins were once united. In this identification we are presented with an amazing picture of the former extent of our carboniferous deposits. * * * * *

A survey of all the circumstances involved in the question of the ancient physical geography of the formation, convinces me that it extended, both in that State and Virginia, at least as far to the S. E. as the great valley. * * * * *

Restricting our attention at present, however, to those districts where the main coal series is developed, we meet with the most ample proofs that all the strata in the insulated basins are precisely on the same geological horizon as those of the great basin W. of the mountains. These coal-rocks all repose conformably on the same easily-recognized formation, the great coal-conglomerate, with the upper beds of which the lower seams are very generally interstratified. * * * * *

Here then we have a coal-formation which, before its original limits were reduced, measured, at a reasonable calculation, 900 miles in length, and in some places more than 200 miles in breadth. I would ask, is it conceivable that any lake, bay, or estuary could have been the receptacle of a deposit so extended, or that any river or rivers could have possessed a delta so vast? The ancient Appalachian ocean grew deeper, as I shall show, towards the W. or N. W.; and inasmuch as rivers push their deltal deposits seawards, and not laterally, and as the carboniferous sediments here to be described are traceable coastwise, as respects this ancient sea, for a length of 900 miles, it is inconceivable how any local fluviatile currents could have assembled them.

Variety of Coal Measures.

Assuming it as susceptible of demonstration, that the various coal-basins, bituminous and anthracitic, of Pennsylvania, Ohio, Maryland, Virginia, Kentucky and Tennessee, were originally united, we may consider the whole as one great formation, in which some highly-interesting gradations in the type and composition of the beds may be traced. To call attention to these phenomena of variation is indeed my main object at this time, since by them only can we arrive at a true theory of the conditions under which the whole were formed. A comprehensive classification of the strata shows the following principal varieties:—

1. Rocks of mechanical origin, of every grade of coarseness, from the smoothest fire-clay to exceedingly rough silicious conglomerates, the whole including within these extremes a wide variety of shales, marls, argillaceous sandstones, and quartzose grits.

2. Limestones, both pure and magnesian, in strata of all thicknesses, from thin bands and narrow layers of detached nodules, to beds measuring from 50 to 100 feet depth. Some of the limestones contain a considerable amount of argillaceous and silicious matter, and many of the thicker deposits consist

of alternating layers of limestone and soft shale. Though a few of these calcareous strata are remarkably destitute of fossils, they are rarely found to be altogether deficient in organic remains, when widely and diligently searched; and some of them quite abound in them. It is especially deserving of note, that the genera are such as invariably indicate oceanic habits. This fact is of the more importance, since some of the limestones occur in immediate contact with beds of coal, and with shales and other strata containing the remains of terrestrial vegetation.

Besides the strata of limestone, we meet with other chemically-formed deposits, in the form of numerous seams of carbonate of iron, and a few considerable beds of regularly stratified chert. The nodular variety of the iron ore is usually imbedded in shale, and lies oftenest adjacent to the coal, while the ore in bands occurs more frequently in contact with the limestone.

3. Coal, in nearly all its known varieties, including every description, from the dryest and most compact anthracites to the most fusible and bituminous kinds of common coal.

Such are the three great classes of strata comprised within the Appalachian coal regions of the United States. If we direct our attention to the manner of their distribution, we shall behold some striking and instructive phenomena, susceptible of reduction to regular and harmonious laws of gradation.

Comparing, in the first place, the rocks of mechanical origin as they occur in different districts, we almost invariably find them coarsest and most massive towards the S.E., and more and more fine-grained and less arenaceous as we pursue them across the successive parallel basins N.W. Thus in the anthracite coal-fields, which are the most South-eastern of all, the coal is interstratified with a vast thickness of rough and ponderous grits, and coarse silicious conglomerates, but is associated with comparatively very little soft clay-slate or shale. In this region the coal-slates themselves are more than ordinarily arenaceous, and bear a smaller proportion to the sandstones than in the basins more to the W. At the same time that the coal-rocks, viewed in the aggregate, acquire a finer texture in going W., the individual strata undergo a corresponding reduction in thickness, while many of them entirely thin away. I may cite, as a striking instance of these changes, the great coal-conglomerate itself, which forms the general base of the main or upper Coal-measures. This massive rock is chiefly composed of large quartzose pebbles imbedded in coarse sand. Adjacent to its most S.E. outcrop in Pennsylvania — that is to say, in Sharp Mountain, where it constitutes the boundary of the first or Pottsville basin—it has a thickness of nearly 1,500 feet; but in the mountains which embrace the Wyoming coal-field, about 30 miles to the N.W., the thickness of the formation is only about 500 feet; while still farther across the chain, where it becomes the general floor of the Coal-measures under the bituminous form, in the basins N.W. of the Allegheny Mountain, its entire thickness seldom exceeds 80 or 100 feet. Tracing it across the great Western coal-field, until we encounter its last outcrop in Western Pennsylvania, Ohio, and Kentucky, this wonderfully-expanded rock dwindles to a thin bed of sandstone, sprinkled with a few pebbles, its whole thickness amounting generally to only 20 or 30 feet. There is a corresponding and quite as striking a diminution in its constituent fragments, the pebbles in the most S. E. belts of the formation being often as large as a hen's egg, while in the N.W. their diameter is reduced to that of a pea.

A similar gradation obtains in the thickness and coarseness of nearly all the interstratified sandstones and other mechanical members of the formation. I conceive that this interesting fact, fully established by the surveys of Pennsylvania and Virginia, shows beyond all question that the S.E. was the quarter whence the coarser materials of the coal-rocks were derived. But there are not wanting other proofs that the ancient land lay in that direction; these will be presently detailed in describing the gradations witnessed in the limestones and beds of coal. The above general law of distribution relates, it should be observed, only to the coarser mechanical aggregates, since there are some apparent exceptions to its generality among the finer-grained slates and shales. Though the texture of these continues to grow finer as we advance W., some of the strata, when individually traced, seem to increase for a certain distance in thickness. This curious circumstance, which belongs indeed to many of the more argillaceous members of our Appalachain formations, so far from invalidating the above inferences respecting the W. transportation of the sediments, come beautifully to confirm them, since it is evident, that until a current, holding in suspension a quantity of sedimentary matter, declines in velocity to a certain point, it cannot let fall any considerable amount of the smaller particles. After it has reached a given degree of retardation, the finer materials will subside, and in an increasing quantity, up to a certain point, at which the loss of velocity in the current is compensated by the exhaustion of material, when a gradual and final thinning of the deposit will take place.

If we examine, in the next place, the gradations of thickness visible in the limestones and other marine deposits, they will be found to lead to precisely similar inferences respecting the position of the ancient land. Viewed either together or individually, the limestones of the Coal-measures of Pennsylvania, Virginia, and Ohio, display a remarkably uniform augmentation as we trace them W. Thus, throughout all the S. E. basins, comprising the whole of the anthracite coal-fields of Pennsylvania, and the Broad Top Mountain in the same State, the formation exhibits a total absence of limestone, and a corresponding deficiency of calcareous matter in the shales and the iron ores. Advancing, however, a distance of 25 or 50 miles N.W. to the general S.E. margin of the great bituminous region, where we enter on the first of the chain of partially-insulated troughs adjacent to the escarpment of the Allegheny Mountain, we no longer encounter a total poverty of limestone, though we still meet with a striking deficiency. As an evidence of this, let us take one of the basins of the Allegheny Mountain—that, for example, which lies near the head of the Potomac River. The minute researches there made in connection with the geological surveys of Virginia and Pennsylvania, have shown that the total thickness of the limestones, counting all the thin bands and layers of nodules, does not probably exceed 10 feet. This statement is confirmed by a pamphlet on the same coal-field, describing the land of the George's Creek Company, by Messrs. Alexander and Tyson. In their very full section of the strata, we do not see a single band of limestone introduced.

Turning to the Moshannon basin in Centre county, which is also a marginal trough of the great W. coal-field, the entire quantity of limestone appears to be about seven or eight feet. If, however, we pass W. from this S. E. line, and cross the great coal-field by any section between the Susquehanna in Pennsylvania and the Little Kenawha in Virginia, we witness a regularly progressive expansion of the calcareous rocks.*

In the upper coal group, or that part of the series which commences with the Pittsburg Seam, the total thickness of pure limestone, excluding numerous thin bands associated with some of the layers of shale, is not less than 150 feet.† Some of the limestone strata of the Coal-measures possess, as will be seen from the second of these tables, a remarkably wide distribution, ranging without interruption from the vicinity of the Allegheny Mountain to the country W. of the Allegheny River. * * * * * Many of these beds of limestone have been traced continuously from Northern Pennsylvania to the Kenawha, and from the E. outcrop, near the Allegheny Mountain, to their W. boundary in Ohio. The marine character of their genera—*Terebratula*, *Goniatites*, *Bellerophon*, *Encrinus*, *&c.*—sufficiently proves that these rocks were originally deposited beneath the waters of an ocean, while, at the same time, the increasing purity of the limestones and the multiplication and expansion of the beds Westward, clearly show that the ancient ocean augmented regularly in depth in that direction. This conclusion, it will be observed, agrees strictly with the results before deduced from the general gradation, visible in the sandstones and other mechanically-formed rocks, which proves that the ancient land was situated towards the E. or S.E.

If we examine the relations of the two classes of the coal strata to each other, the land-derived and sea-derived rocks, we perceive that the latter, or the limestones, thicken, going West, at the expense of the former. Frequently two beds approach, and either entirely coalesce, or remain divided by only a thin marly shale, formed from the residual finely-subdivided matter wafted out by the currents, which, farther E., or nearer the

* Here follows the tabular statement given in the foot-note on page 24 H, of this report.

† See "Report of Geological Survey of Virginia, for 1840.'

land, deposited the coarser and thicker sandstones and arenaceous slates.

While this gradation shows itself, new beds of calcareous rock interpolate themselves in new positions in the series, and many of the sandstones thin away and cease altogether, so that the whole formation becomes, by both these changes, more and more oceanic in its type. But the most important result of this mode of tracing the strata is the evidence we have of the frequent alternation of a tranquil and disturbed condition of the waters. * * * * *

With these considerations before us, we cannot fail to perceive in the Appalachian coal strata, the monuments of many alternate periods of movement and total or comparative rest. If it be conceded that each of the purer beds of limestone, remarkable for the extreme fineness of their texture and the absence of foreign sedimentary matter, is the index of a longer or a shorter interval of tranquility in the waters, we shall discern (omitting for the present all similar inferences to be derived from the coal-seams) a much greater number of such separate periods than a mere enumeration of the individual beds would indicate, unless we count the interstratified shales and marls. These last-mentioned strata, generally assuming, as we go E., a thicker and coarser type, furnish as unequivocal a record of disturbances as if the spaces they occupy between the beds of limestone were filled by the coarsest mechanical aggregates.

One of the most interesting general questions connected with the land and sea produced strata relates to the physical geography of the ancient coast near to which they were deposited, and the inquiry at once suggests itself, whether the receptacle of these various sediments was an extensive estuary receiving the silts of some gigantic river or rivers, or a vast expanse of shallow sea, bounded by a long line of coast, upon which the successive deposits were formed by a very different agency from any we can ascribe to ordinary fluviatile or littoral currents.
* * * * *

Geographical Range of Coal Beds.

Of the facts connected with the range of the individual coal-seams, that of their prodigious extent is itself one of the most surprising and instructive. As a general rule, this wide expansion characterizes all the beds of both the bituminous and anthracite basins. It is true that many seams possess a comparatively local range, but not a few of those which, on first examination, appear of circumscribed extent, cover in reality, a very wide area, the error respecting them being caused by fluctuations of thickness, or by their occasionally thinning out and reappearing. Among those which manifest great permanency as to thickness, the vast range of some of the larger ones is truly extraordinary. Let us trace, for example, the great bed which occurs so finely exposed at Pittsburg, and along nearly the whole length of the Monongahela river, and which I have called the Pittsburg seam. * * * * * The limits of this bed, as at present known, [1858,] are nearly as follows: That portion, by far the largest part, which is contained in the great Western basin, has its N. termination in Indiana County, in Pennsylvania, and its S. W. on the Ohio River below Guyandotte. The general S. E. outcrop ranges along the W. foot of Chestnut Ridge, or West Laurel Hill, from Indiana County to Tygart's River, in Virginia. * * * * * The longest diameter of the great elliptical area here delineated is very nearly 225 miles, and its maximum breadth about 100 miles. The superficial extent of the whole coal-seam, as nearly as I can estimate it, is about 14,000 square miles.

But the limits here described, though wide, fall very far within those which the bed anciently occupied. To the S. E. of the large basin of the Ohio River there are several other insulated parallel troughs, which also contain the Pittsburg Seam. Of these, the farthest from the main coal-field is that at the head of the Potomac River, at a distance of about 43 miles in a straight line. The E. margin of the Pittsburg Bed is here, however, nearly 50 miles E. S. E. of the E. edge of the same seam in the main or W. basin, and it has a corresponding expansion E. in other districts. That this coal-bed preserves an unbroken range for many miles to the N. E. of the termination of the principal basin in Indiana County, appears highly probable, from a comparison of the Coal-measures at certain localities in that quarter. I shall not, however, assume the known length of the tract actually occupied by it as exceeding the above-mentioned 225 miles, throughout which it is uninterruptedly traceable. If we now take into account the 50 additional miles of breadth which the bed once possessed, its former area must have been at least 34,000 square miles, a superficial extent greater than that of Scotland or Ireland.

Though the above is perhaps the greatest extent of surface which it is in our power positively to assign to this bed of coal, the proofs of a prodigious denudation of the strata, throughout the districts bordering its present outcrop, are so irresistible that I consider the dimensions here given as bearing actually but a small proportion to the real ancient limits of the stratum.

But restricting our attention for the present to those limits which it did undoubtedly once occupy, it is still by far the most extensive coal-bed yet explored in any country; and the mere fact of its great extent must exert an influence on our views concerning the conditions under which the whole coal-formation originated.

The general uniformity in the thickness of this superb bed, throughout so vast a region, and at the same time the regular and gentle gradation which it experiences in size when we trace it from one outcrop to the other, are features not less remarkable than its enormous length and breadth. * * * * *

While we are thus furnished with conclusive evidence, from the fact that its rate of increase is most rapid towards the S. E., that the ancient land with which the stratum was connected must have been situated in that direction, we see that the N. E. part of the coast was the quarter where its materials were supplied in the greatest abundance. To this conclusion I am disposed to appeal, in support of the conjecture already ventured, that this great bed of the main or W. coal-field is but a remnant of a still more expanded stratum, which attained its maximum size in the enormous seam of which all the anthracite basins present us insulated patches. The singular constancy in the thickness of this Pittsburg Bed, no less than its prodigious range, are circumstances that seem strongly adverse to the theory which ascribes the formation of such deposits to any species of *drifting* action. * * * * *

Intimate Mechanical Structure of Coal.

An examination of the structure of the coal itself, apart from the fact of the great range and uniformity in the thickness of the beds, renders it apparent that no irregular dispersion of the vegetable matter by any conceivable mode of drifting, either into estuaries or the open sea, could produce the phenomena which they exhibit. The mechanical arrangement of the layers in every coal-seam, as seen when viewed edgewise, indicates plainly that it is a compound stratum as much as any other sedimentary deposit, each bed being made up of innumerable very thin laminæ of glossy coal, alternating with equally minute plates of impure coal, containing a small admixture of finely-divided earthy matter. These subdivisions, differing in their lustre and fracture, are frequently of excessive thinness, the less brilliant leaves sometimes not exceeding the thickness of a sheet of paper. In many of the purer coal-beds, both anthracitic and bituminous, these thin partings between the more lustrous layers consist of little laminæ of pure fibrous charcoal, in which we may discover the peculiar texture of the leaves, fronds, and even the bark of the plants which supplied a part of the vegetable matter of the bed. If traced out to their edges, all these ultimate divisions of a mass of coal will be found to extend over a surprisingly large surface when we consider their minute thickness. Pursuing any given brilliant layer, whose thickness may not exceed the fourth part of an inch, we may observe it to extend over a superficial space which is wholly incompatible with the idea that it can have been derived from the flattened trunk or limb of any arborescent plant, however compressible. When a very large block of coal is thus minutely and carefully dissected, it very seldom, if ever, gives the slightest evidence of having been produced from the more solid parts of trees, though it may abound in fragments of their fronds and deciduous extremities. The laminæ of brilliant carbonaceous matter almost invariably thin away to a fine edge before they terminate, a fact which of itself seems to prove that the material was in a soft or pulpy state at the time of its accumulation, and this supposition receives countenance from the homogeneous texture and conchoidal fracture of every such layer.

Granting the correctness of this inference, which is not in conflict with the beautiful microscopic determinations by Hutton respecting the traces of vegetable structure in certain portions of coal, the argument seem almost conclusive that the vegetable matter grew where it was deposited. * * * * *

Though this observer found more or less of the cellular vegetable structure in each of the three varities of Newcastle coal, he discovered a complete obliteration of the characteristic cells in those finest lustrous portions of the caking coal, where the crystalline structure, as he terms it, is best developed.

Besides the above-mentioned features, all the coal beds which I have ever examined, or seen minutely described, possess another peculiarity in their mechanical constitution, on a less minute scale, which is equally incompatible with the notion of transportation by currents.

I refer here to the subordinate divisions of the coal-beds, some of which are strata of pure coal, some of earthy coal, and some of common shale, all constituting together the compound mass which we call a coal-seam, but each maintaining its particular position and character as a distinct deposit over an area which is truly astonishing. Those persons who are conversant with large mining districts are aware of the many instances of remarkable persistency in these subdivisions in the coal-beds, since it is frequently by their means that the miners recognize a known coal-seam in cases of difficulty.

Thus the largest bed of the anthracite fields of Pennsylvania contains almost everywhere a particular band of unusually pure coal, not far from the bottom, generally from 3 to 4 feet in thickness.

A still more striking example occurs in the great Pittsburg Bed already spoken of. If we dissect this compound mass, and trace the several divisions, we become impressed with the wonderful distances to which some of them extend. Not to enter here into a minute discussion of all the features of this widely-distributed seam, it will suffice to state that it consists principally of three members which are readily recognized. The lowest is a thick bed of uncommonly pure coal, the middle a layer of soft shale or fire-clay, about 1 foot in thickness, and the uppermost, or roof coal, is itself a compound seam, 2 or 3 feet thick, of alternating layers of coal and fire-clay. Now it is a highly-instructive fact, that this general triple subdivision prevails throughout nearly the whole range of the seam from its E. to its W. outcrops, and from the Conemaugh in Pennsylvania into Western Virginia, for a distance of more than 200 miles, from N.E. to S.W.

But besides this fact, each subordinate portion preserves its own distinctive features, the upper member being everywhere remarkable for its alternation of thin bands of coal and shale. Can any evidence be more conclusive as to the uniformity ot the conditions under which every part of this coal-bed was produced?

There must, indeed, have prevailed an almost perfect uniformity in the state of the surface throughout the vast area which it occupies, as respects even the formation of the thinnest of these subdivisions. Such remarkable sameness of action throughout the same geological horizon, appears absolutely incompatible with any mode of drifting of the vegetable matter.

Only one particular process of accumulation appears to explain the occurrence, in such cases, of these thin and uniform sheets of material, of which the thickness is often less than a foot, while their superficial area is many hundred square miles. I cannot conceive any state of the surface adopted to account for these appearances, but that in which the margin of the sea was occupied by vast marine savannahs of some peat-creating plant, growing half-immersed on a perfectly horizontal plane, and this fringed and interspersed with forests of trees, shedding their leaves upon the marsh. Such are the only circumstances under which it is likely that these regularly parallel, thin, and widely-extended sheets of carbonaceous matter, could have been accumulated.

Independently of the above argument, based on the breadth and uniform distribution of the layers in the coal there is another, drawn from the striking deficiency of earthy sedimentary particles. In many of the purest layers, the total proportion by weight of foreign mineral substance in the coal, is less than two per cent., sometimes barely one per cent., while the ratio by bulk is consequently less than one-half of this. So extremely insignificant a quantity is what we should expect, on the hypothesis of a tranquil accumulation in wide sea-meadows, extending far out from the edge of the ancient shore, where no turbid currents could get access. It is as inconsistent, on the other hand, with the notion of a drifting of the vegetable matter itself, which, according to any conceivable mode of transportation, would be accompanied by a large amount of earthy matter, such as abounds in all deltal deposits, and even mingles with the wood in the raft of the Atchafalaya. That so nearly the whole of the suspended mineral matter, even to the fine particles of impalpable clay, should have subsided, in almost every instance, before the first portions of the floating vegetation sank, contradicts all observation respecting similar actions now occurring. The introduction of any argillaceous matter into the transparent waters of the great peat-morasses, must have happened in the manner of an exceedingly quiet and diffused silting in, or more properly a slow intermingling, of very slightly turbid water with that of the limpid sea. The above arguments from the uniformity in the distribution of the vegetable matter of the coal-seams, and from the absence of earthy matters in the coal, have been already employed by Mr. Beaumont as objections to the drift theory, in a communication read to the Geological Society of London, February 26, 1840.*

Floor of the Coal.

The deposit upon which each seam of coal immediately rests, and which I shall call the floor, is, with a few rare exceptions, wholly distinct in its composition from the roof, or that which reposes directly upon the bed. To Sir William Logan we are indebted for having ascertained the highly important fact, that the floor of every coal-seam in South Wales is composed of a peculiar variety of more or less sandy clay, distinguished by its containing the *Stigmaria ficoides*. "Portions of the stem of the *Stigmaria* are found in other parts of the Coal-measures, but it is only in the under clay that the fibrous processes are attached to the stem, or associated with it."† * * * * * "When it is considered that, over so considerable an area as the coal-field of South Wales, not a seam has been discovered without an under-clay abounding in Stigmaria, it is impossible to avoid the inference, that there is some essential and necessary connection between the existence of the Stigmaria and the

*BEAUMONT, "Proceedings of the Geological Society," No. 69

†LOGAN, *Ibid.*

production of the coal. To account for their unfailing combination by drift seems unsatisfactory; but whatever may be the mutual dependence of the phenomena, it affords reasonable grounds to suppose that the *Stigmaria ficoides* is the plant to which we may mainly ascribe the vast stores of fossil fuel."†

* * * * *

Roof of the Coal.

If we examine, in the next place, the strata which immediately rest upon the coal, we shall discover a condition of things in striking contrast with the phenomena of the under-clay. Instead of one uniform material almost invariably present, composed of finely-divided particles, the beds overlying the coal consist of nearly every variety of rock embraced in the formation, though they are more usually some form of laminated carbonaceous slate. Both in its composition and structure, the roof rock manifests signs of having been deposited by a more or less rapid current. In place of a single species of fossil plant, it usually includes a prodigious variety; and the delicate ramifications of these, instead of intersecting the bed in various directions, as the processes of the Stigmaria do in the fire-clay, lie in a singularly disordered and fragmentary condition, in planes almost invariably parallel to the bedding. Lindley and Hutton, in their work on the *Fossil Flora of Great Britain*, give the following very accurate description of the mode in which the organic remains occur in the roof-slates in England, and the account is equally applicable to those of the United States: "It is the beds of shale, or argillaceous schistus, which afford the most abundant supply of these curious relics of a former world, the fine particles of which they are composed having sealed up and retained in wonderful perfection and beauty the most delicate forms of the vegetable organic structure. Where shale forms the roof of the workable seams of coal, as it generally does, we have the most abundant display of fossils. The

† Mr. Rogers then gives credit to Mr. Edward Mammatt for anticipating Mr. Logan's discovery, in his book on the Coal-Field of Ashby de la Zouch, 1834; and also to De Luc, M'Culloch, Jameson, Brogniart, Lindley and others, for holding the same views. Brogniart and Binney supposed that they proved that the *Stigmaria* was the root of Sigillaria. Lesquereux has expressed an opinion that the Sigillaria was the flower stalk of the water-plant Stigmaria, the "rootlets" of which were its *leaves*.

principal deposit is not in immediate contact with the coal, but from 12 to 20 inches above it;* and such is the immense profusion in this situation, that they are not unfrequently the cause of very serious accidents, by breaking the adhesion of the shale-bed, and causing it to separate and fall, when, by the operation of the miner, the coal which supported it is removed. After an extensive fall of this kind has taken place, it is a curious sight to see the mine covered with these vegetable forms, some of them of great beauty and delicacy; *and the observer cannot fail to be struck with the extraordinary confusion, and the numerous marks of strong mechanical action, exhibited by their broken and disjointed remains.*" Such is the nature of the roof when it consists of the usual carbonaceous shale or slate; but in the Appalachian coal-fields it is oftentimes a much coarser rock, being either an argillaceous flaggy sandstone, or a coarse arenaceous grit, or even occasionally a silicious conglomerate. In these instances the enclosed vegetable remains are for the most part fragments of the larger stems or branches of gigantic aborescent plants, their fronds and leaves being less abundant. These fragments occur in all postures as respects the plane of the bedding—horizontally, obliquely, or perpendicularly; and betray, in their broken condition and irregular mode of dispersion, the sudden and tempestuous character of the currents which drifted and entombed them. Though the arenaceous rocks having these features sometimes rest in immediate contact with the upper surfaces of the beds of coal, they more frequently lie at a moderate distance over them, an argillaceous laminated slate interposing to form the actual roof. A further indication of the violence of the currents which strewed these coarse materials over the coal is sometimes to be detected in the composition of the lowest portion of the overlying bed of grit or sandstone, in which a large amount of coal, in the state of powder or sand, is disseminated in the rock, giving it a dark speckled appearance. This is of very common occurrence in the anthracite coal-strata of Pennsylvania, where the coarse grit not unfrequently rests immediately on the coal. It implies, I conceive, the erosion of a certain portion of the upper surface of the soft carbonaceous mass by the friction of the sandy cur-

*This refers to the special coal beds examined by them.—J. P. L.

rent.* The coaly matter, thus disturbed, would subside with the first layers of the sand with which it was mingled. Sir William Logan has mentioned a still more striking proof of the energy of the movements which occasionally occurred during the formation of the Coal-measures. He gives an account of actual boulders, or rounded pebbles of coal, in the Pennant grit, and other coarse strata of the coal-field of South Wales.† * * *

Direct Contact of Coal-Beds and Marine Limestones.

As the actual contact of beds of coal and limestone is of rare occurrence in the coal-fields of other countries, and as the circumstance must have an influential bearing on all our speculations concerning the physical conditions prevailing at the formation of the strata, and, to a certain extent, on our whole theory of the origin of coal, I shall here describe some of the best-known instances before I reason concerning them. * * *

First. In the lower division of the main Coal-measures there occurs, near the town of Mercer, in Pennsylvania, a seam of good coal, having a thickness of about 2 feet, which is immediately overlaid by a bed of very pure limestone, also about 2 feet thick, containing a variety of marine organic remains of the genera *Terebratula*, *Bellerophon*, &c. In some spots the pure coal is not separated from the pure limestone by more than a single inch, or at most 2 inches, and then the interval is filled with a calcareo-carbonaceous shale.

Secondly. Higher in the series, but still in the lower part of the main Coal-measures of Western Pennsylvania, we meet with a bed of fossiliferous limestone, the thickness of which, in many neighborhoods near the Allegheny River, is about 15 feet. It contains several oceanic species, among them some *Crinoideæ*, two species of *Terebratula*, and a *Goniatites*. In some places this stratum embraces a thin seam of coal, 4 inches thick, in almost direct contact with the limestone.

Thirdly. The limestone, which is the first underneath the Pittsburg Seam, contains a bed of coal 1 foot in thickness, separating two of its lower layers.

* The upper part of the Coal bed is usually absent (by erosion) whenever the roof is coarse sandstone.—J. P. L.

† See such an instance found in Clearfield county, and recorded on page 27 H, above.

Fourthly. Near Pittsburg, the great coal-seam frequently rests within a few inches of this underlying limestone, in which are a few occasional fossils, all of marine genera. In these places the dividing layer is only a few inches thick, and consists of a bluish fire-clay.

Fifthly. In Fayette County, Pennsylvania, the great limestone, which lies above the Pittsburg coal-bed, encloses very generally two thin seams of perfect coal, immediately in contact with the layers of the rock. These coals appear to have a considerable range, extending into the adjoining counties. The largest is occasionally 2½ feet thick, and a few inches of black calcareous slate alone separate it from the hard limestone. The other coal-bed has a thickness of about 1 foot, and its surfaces are in equally close contact with the limestone. Neither of these beds is as widely expanded as the including limestone.

Sixthly. Underneath the uppermost workable bed of coal in Western Pennsylvania, or that which I have termed in my Reports the Waynesburg Seam, there is a stratum of limestone, which sometimes encloses a thin coal-bed, measuring about 1 foot.

Seventhly. At Putnam Hill, near Zanesville, in Ohio, a bed of limestone, 5 feet thick, rests, according to Dr. Hildreth, on a seam of coal 1 foot thick, there not being more than 2 inches of fire-clay interposed. The limestone contains *Encrini*, *Terebratulæ*, and other marine fossils.*

Eighthly. The same writer mentions that, on the Clear Fork of Little Muskingum, in Ohio, there is a seam of good bituminous coal, 3 feet thick, reposing directly on a dark, carbonaceous, fossiliferous limestone 8 feet in thickness. It is overlaid by another limestone, measuring 6 feet, from which it is separated by a very thin layer of shale.

Ninthly. Dr. Hildreth further states, that on Wills's Creek, in the same region, a coal-seam 5 feet thick occurs, resting immediately on a bed of limestone, the thickness of which is 20 feet.

I might cite a large additional number of cases in Pennsylvania, Virginia and Ohio, in proof of the frequency of the contact of the coal-seams and beds of limestone; but I have been

* Hildreth, in *American Journal of Science*, p. 31

disposed to establish the fact chiefly from instances in different portions of the formation, and to show that the contiguity of the coal and limestone is often maintained throughout a considerable extent of country.*

Theory of the origin of the Coal Strata.

M. Adolphe Brongniart, after comparing the phenomena of the ancient coal and its fossil plants with the great peat-mosses of the present day, states in a memoir published in 1838, that he continues to adhere to the opinions originally advanced by Werner and De Luc, that the vegetation entombed in the carboniferous strata chiefly grew in the localities where the coal is now found.† * * * * *

Mr. Hawkshaw, in a communication to the Geological Society of London in 1839, describes the remarkable phenomenon of five fossil trees, exposed in a cutting on the Manchester and Bolton railway, standing erect in relation to a bed of coal, 8 or 10 inches thick, and in the same place with their roots. The largest of these was five feet in diameter at the base, and eleven feet high. He conceives it probable that they grew where they occur. ‡ * * * * * "If the coal be considered as the debris of a forest, it is difficult to account for not finding more trunks of trees than have been discovered in our coal basins; and it is only, perhaps, by allowing the original of our coal-seams to have been a combination of vegetable matter analogous to peat, that the difficulty can be solved. § * * * * *

Mr. Beaumont, in a paper read to the same Society, November 6, 1838, * * * * * states that the vegetation which formed the coal grew on swampy islands, that it consisted of *ferns*, *calamites*, *coniferous trees*, &c., which operated, through their decay and regeneration, to form peat-bogs; and that the islands, by subsiding, were covered with drifted sand, clay, and shells, till they again became dry land, and supported

* A very remarkable instance is seen in the Deep River Coal Bed of North Carolina, which is divided into two benches or thick layers of coal, separated by a thick layer of carbonate of iron, holding fish teeth.—J. P. L.

† LYELL'S *Elements*, ii, 135, Boston edition.

‡ HAWKSHAW, in "Proceedings of the Geological Society, London," No. 64.

§ *Ibid.*, No. 69.

another vegetation; and this process, he supposes, was repeated as often as there are coal seams.*

Dr. Buckland, * * * * * observes, * * * * that some of the trees which are found erect in the coal formation have not been drifted, is, I think, established on sufficient evidence; but there is equal evidence to show that other trees and leaves innumerable, which pervade the strata that alternate with the coal, have been removed by water to considerable distances from the spots on which they grew. Proofs are daily increasing in favor of both opinions, namely, that some of the vegetables which forms our beds of coal grew on the identical banks of sand and silt and mud, which, being now indurated to stone and shale, form the strata that accompany the coal; whilst other portions of these plants have been drifted to various distances from the swamps, savannahs, and forests, that gave them birth; particularly those that are dispersed through the sandstones, or mixed with fishes in the shale beds. * * * * *

Another paper on the same subject of the fossil trees, found on the Manchester and Bolton Railway, was read contemporaneously with the last communication of Mr. Hawkshaw. The author, Mr. Bowman, is of opinion "that * * * * * the coal has been formed from plants which grew on the areas now occupied by the seams; that each successive race of vegetation was gradually submerged beneath the level of the water, and covered up by sediment, which accumulated till it formed another dry surface for the growth of another series of *trees* and plants, and that the submergences and accumulations took place as many times as there are seams of coal.

In reviewing the above facts and opinions, Dr. Buckland conceives that a luxuriant growth of marsh plants, as *Calamites*, *Lepidodendra*, *Sigillaria*, &c., may have formed a superstratum of coal, resting on a substratum of the same, composed exclusively of remains of *Stigmariæ;* and in accounting for the marine and fresh-water strata alternating with the coal-beds, he appeals to the intermitting and alternate processes of subsidence, drift, and vegetable growth.† * * * * *

I may now venture to advance my own explanation of the

*BEAUMONT, in "Proceedings of the Geological Society, London," No. 65.
†Anniversary address to Geological Society, 1841.

phenomena, and to indicate wherein I differ from the able authors I have cited. * * * * *

Let us imagine the areas now covered with the coal-formation, to have possessed a physical geography, in which the principal feature was the existence of extensive flats, bordering a continent, and forming the shores of an ocean, or some vast bay, outside of which was a wide expanse of shallow but open sea. Let us now suppose that the whole period of the Coal-measures was characterized by a general slow subsidence of these coasts, on which we conceive that the vegetation of the coal grew;—that this vertical depression was, however, interrupted by pauses and gradual upward movements of less frequency and duration, and that these nearly statical conditions of the land alternated with great paroxysmal displacements of the level, caused by those mighty pulsations of the crust which we call earthquakes. Let us further conceive, that during the periods of gentle depression, or almost absolute rest, the low coast was fringed by great marshy tracts or peat-bogs, derived from and supporting a luxuriant growth of *Stigmariæ* (*Sigillaria* and *Lepidodendra*,) and that along the landward margin, and in the drier places of these extensive sea morasses, grew the *Coniferæ*, *Tree-Ferns*, *Lycopodiæceæ*, and other arborescent plants, whose remains are so profusely scattered throughout the coarser strata between the coal seams. In this condition of things, the constant decomposition and growth of the meadows of *Stigmariæ* would produce a very uniform, extended stratum of pulpy but minutely-laminated pure peat. This would receive occasional contributions from the droppings by the scattered trees of their leaves, fronds, and smaller portions, which, being driven by winds or floated on the high tides, would lodge among the Stigmariæ in the marshes, and slightly augment the deposit. These leaves and fronds, covered over more or less rapidly by the growing *Stigmariæ*, or varying in their tendency to decay, according to the abundance or deficiency of their juices, would, when thus enclosed, pass at once either to the pulpy state, and ultimately form coal, or, by the more rapid extrication of their volatile portions, remain as mineral charcoal, and preserve their vegetable fibrous structure. In both of these conditions of coal and charcoal, we often find the smaller parts of plants retaining their organized forms among the laminæ of the purest coal-seams. Upon this view of a gradual accumulation from the *Stigmariæ*, assisted by the deciduous parts of the trees, it is altogether unnecessary to suppose that any portions, even the upper layers of the coal-beds, derived their vegetable matter from the stems of the trees themselves. Thus the absence of trunks and roots from coal is reconciled with the occasional occurrence of their fronds and lighter extremities. Upon no other hypothesis respecting the physical condition of the region which produced the coal vegetation than that here imagined, can I explain the singular infrequency of fossil trunks standing on or in the coal, or account for their occasional occurrence. as in the instances described by Hawkshaw and Bowman. No other supposition seems to furnish a cause for the absence of all traces in the coal itself of the larger parts of arborescent plants. and for their equally remarkable abundance in a broken and dispersed state in the overlying strata.

Assuming such to have been the condition of the surface during the tranquil periods of the accumulation of each coal-bed. we may conceive the other strata to have been produced in the following manner. Let us suppose an earthquake, possessing the characteristic undulatory movement of the crust, in which I believe all earthquakes essentially to consist, suddenly to have disturbed the level of the wide peat-morasses and adjoining flat tracts of forest on the one side, and the shallow sea on the other. The ocean, as usual in earthquakes, would drain off its waters for a moment from the great Stigmaria marsh, and from all the swampy forests which skirted it, and, by its recession, stir up the muddy soil, and drift away the fronds, twigs, and smaller plants, and spread these and the mud broadly over the surface of the bog. In this way may have been formed the laminated slates, so full of fragmentary leaves and twigs, which generally compose the immediate covering of the coal-beds. Presently, however, the sea would roll in with impetuous force, and reaching the fast land, prostrate everything before it. Almost the entire forest would be uprooted, and borne off on its tremendous surf. Spreading far inland, compared with its accustomed shore, it would wash up the soil, and abrade whatever fragmentary materials lay in its path, and, loaded with these, it would then rush out again, with irresistible violence, towards its deeper bed, strewing the products of the land in a coarse promiscuous stratum, embedding the fragments of the broken and disordered trees. Alternately swelling and retiring with a suddenness and energy far surpassing that of any tide, and maintained probably in this state of tempestuous oscillation by fresh heavings of the crust, the waters would go on spreading a succession of coarser or finer strata, and entombing at each inundation a new portion of the floating forest. Upon the dying away of the earthquake undulations, the sea, once more restored to tranquillity, would hold in suspension at last only the most finely-subdivided sedimentary matter, and the most buoyant of the uptorn vegetation—that is to say, the argillaceous particles of the fire-clay—and the naturally floating stems of the plants. These would at last precipitate themselves together by a slow subsidence, and form a uniform deposit, exhibiting but few traces of any active horizontal currents, such as would arise from a drifting into the sea from rivers. The chief portion of the coarser fire-clay would settle first, and then the more impalpable particles, in company with the stems and leaves of the uprooted vegetation. Thus we may account for the constant reproduction of the peculiar soil of the coal-seams, and for the preservation, particularly in its upper clayey layers, of the Stigmaria (Sigillaria); the simple consequence of the final subsidence of these materials being the production of the necessary substratum of another coal marsh. The marine savannahs becoming again clothed with their matting of vegetation, and fringed on the side towards the land with wet forests of arborescent ferns and other trees, all the essential conditions and changes that constituted this wonderful cycle in the statical and dynamic processes belonging to each seam of coal, and the beds enclosing it, would be completed, and ready to be once more renewed.

Though the train of actions here imagined enables us to reconcile the indications afforded by the coal-beds of periods of prolonged tranquillity, with the evidences of violent aqueous currents, as shown in the composition of the coarser mechanical rocks; yet a complete theory of the coal-formation calls for the introduction of other considerations connected with the existence and positions of strata derived from chemical and organic agen-

...ies, as the limestones, cherts, and beds of carbonate of iron. * * * * * Sir Charles Lyell, in an early edition of his *Elements of Geology*, says: * * * * * "if we appreciate the full strength of the evidence in favor of continued subsidence in the coal-field of South Wales, we shall be the less surprised to learn that the vertical depth of the superimposed strata is enormous, amounting in some places to no less than twelve thousand feet."* Though a vast preponderance of subsidence over elevation is plainly indicated in the prodigious thickness of the coal-measures, each particular coal-seam in which, was produced successively at the surface, I cannot conceive that either an alteration of periods of subsidence and repose, or an uninterrupted prolonged depression, will explain the phenomena of the Appalachian coal-rocks, as they have been here described. * * * * * Considering the many striking instances which I have recorded of the close approach or actual contact of certain beds of coal and oceanic limestone, we cannot resist the conclusion, that the gradual downward movement was frequently interrupted by a slow upward one.

In all the incidents that I have cited, where the limestone stratum immediately underlies a coal-seam, it is obvious that an upward movement of the land must have taken place, so gradually as to be unattended by any sensible commotion of the waters. A considerable and sudden lifting of the bed of the sea would infallibly have caused the production of violent currents, competent to spread over the quiet precipitate of limestone, one or more coarse arenaceous or argillo-arenaceous strata.

That the intervals of repose indicated by the limestones were, like those of the beds of coal, sometimes suddenly terminated by earthquake disturbances, strewing over the marine sediments the materials of the land, is manifest from the phenomena, though it is not less clear that the cessation of the periods of relative tranquillity, marked by very gradual subsidence, must in all cases, where the coal-beds are overlaid directly by marine limestones, have been effected by simply a more rapid process of depression. Every superimposed limestone, without an intervening roof-slate or sandstone to separate it from the coal, affords, I conceive, a conclusive proof of this increased rate of submergence.

* LYELL'S *Elements*, Boston edition, vol. ii, pp. 128, 134.

Perhaps it will be objected, that a merely accelerated subsidence, such as I have here supposed, if taken in conjunction with the hypothesis of a drifting of the land materials by rivers, will satisfactorily explain all the facts which I have ascribed to the turbulent movement of the sea against the land during earthquakes.

But though it is highly probable, from the phenomena of nearly every extensive coal-field, that rivers did carry into the parts of the ocean and its estuaries, now drained and occupied by the coal strata, a considerable quantity of argillaceous deltal deposits, yet it is difficult to imagine how any moderately rapid subsidence, if unaccompanied by some *paroxysmal* movement, could create a current energetic enough to uproot and float away nearly the whole of those vast forests, which evidently grew close to the site of each seam of coal, and to snap off to the stumps even the most colossal trees. Nor is it easy to explain why such a quiet submersion of the swampy forests did not result in the preservation of the trees in their original erect posture by the drifting around them of the supposed river sediments.

It is fair to infer, that so long a line of coast as we must conceive bordered the Appalachian ocean, if we may judge from the great longitudinal extent of some of our coal-seams, was not destitute of rivers, and we are therefore constrained to admit that some amount of sedimentary matter must have entered the sea in that manner; but at the same time, we have only to notice the striking deficiency of earthy matter in the numerous coal-beds, and in many of the strata of limestone, to be persuaded that the amount of material contributed to the coal-measures by fluviatile transport was relatively inconsiderable. It may be fairly questioned, whether any sensible proportion of river silt could spread itself to the distance of 150 or 200 miles seawards, over the great coal morasses of such a coast; and yet we are compelled to assume this, if we deny the above paroxysmal theory.

That the geological and geographical changes known to have been caused in modern times by earthquakes, entitle us to speculate upon their agency in suddenly shifting the level of the low tracts once occupied by the marshes and swamps of the coal-seams, must, I think, be conceded. Few geologists will deny the probability of frequent changes, in the carboniferous period, analogous to that which took place in the great plain at the mouth of the Indus in the year 1819, when "extensive flats bordering the Indus sank down, and, for many years after, vessels were forced through the boughs of the tamarisk trees, still standing erect." *

Should the foregoing theory, based on the complicated statical and dynamic phenomena of the Appalachian coal strata, be correct, then have we, in every stratum of the series, not merely a new picture of the physical geography of the region, but a clear legible record of the very changes, gradual or tempestuous, of which each in its turn was the result. We unclasp, as it were, a whole volume of hydrographic charts, displaying, for a vast succession of epochs, the ever-changing relations of the land and waters. A wide tract of ancient coast is at one time occupied by the ocean, at another by an immense plain filled with green marshes and swamps, and at another by dry land clothed with a tangled forest. But we behold more than merely these several conditions of the surface: we perceive the very transitions themselves which revolutionized the geography; we discern the ocean in the very act of encroaching on the land, forming extensive marshes where before the whole was solid shore; we actually trace it in its gradual retreat, exposing its own marine sediments to form a fertile soil for vast savannahs, and again we see the entire region embracing the dry land, the marshes, and the sea, heaving and undulating in the billows of the irresistible earthquake, the ocean and the land contending for mastery in the tremendous conflict.

If the Appalachian coal strata, whose history I have here endeavored to interpret, exhibit truly the above-imagined conditions and events, we may consider the entire formation as constituting a stupendous tide-gauge, registering the lengthened ebbings and flowings of that ancient sea, and the stormy agitations of its oscillating waters, as the epoch of its last greatest movement and final drainage drew near.

Gradation of volatile matter in the Coal Beds.

There prevails a very interesting law of gradation, in the

* LYELL'S *Elements*, vol. ii. p. 136.

quantity of volatile matter belonging to the coal, as we cross the Appalachian basins from the S.E. towards the N.W. The extraordinary extent of area over which this law obtains, and its intimate connection with corresponding gradations in the structural phenomena of the region, the description and theory of which have been given elsewhere by Professor W. B. Rogers and myself, seem to claim for it a place in the present general account of our coal-measures. The gradation may be thus briefly described:—Crossing the Appalachian coal-fields, N.W. from the great valley to the middle of the main or W. trough, by any section between the N.E. termination of the formation in Pennsylvania and the latitude of Tennessee, we find, as the result of multiplied chemical analyses, a progressive increase in the proportion of the volatile matter, passing from a nearly total deficiency of it in the driest anthracites, to an ample abundance in the richest gaseous coals. The existence of this singular law of transition was first ascertained by me in 1837, in which year I made mention of it in some public lectures. It was communicated to the Association of American Geologists, at their first annual meeting, in the spring of 1840; but I did not publish it in print until the following winter, when it was briefly alluded to in my Fifth Annual Report on the Geological Survey of Pennsylvania. Evidence of the existence of such a gradation in the coals of Western Virginia will be found in the annual reports of the Geological Survey of that State for the years 1839 and 1840. These historical references are here introduced, because the determination of the general fact was the result of many laborious analyses of our coals, made by my brother and myself, and because attempts have been made by others to establish a claim to the discovery. The lists of analyses contained in the Reports of the Surveys of Pennsylvania and Virginia, and similar data not yet published, show the following as the general proportion of the bituminous matter in the different belts of the formation, as we cross the region from S.E. to N.W.

First. In the most S.E. chain of basins the coal is, for the most part, a genuine anthracite, containing, however, sometimes a small percentage of bitumen, and always a little gaseous matter, chiefly hydrogen. The quantity of the volatile matter varies, according to geological locality, from about six to twelve or fourteen per cent. This first belt of basins embraces all the anthracite coal-fields of Pennsylvania, the slightly bituminous ones of Broad Top on the Juniata, of Sleepy Creek, of the Little North Mountain, of Catawba Creek, Tom's Creek, Strouble's Run, and Brushy Ridge, in Virginia. The coal of the Little North Mountain is, however, a true anthracite. All of these coal-fields, and insulated patches of the formation, belong to the most disturbed portions of the Appalachian chain, and they are associated with some of the boldest flexures and greatest dislocations of the whole region. The first or S.E. anthracite basin of Pennsylvania presents innumerable sharp flexures and close plications, with inversion, of the strata.

Secondly. In the next well-defined range of basins farther towards the N.W.—that, namely, of the Allegheny Mountain, and the general escarpment of which it is a part—the proportion of volatile matter varies usually from sixteen to twenty-two per cent., but is generally about eighteen or twenty per cent. This belt includes all the coal-fields situated immediately to the N.W. of the Allegheny Mountain in Pennsylvania, also the Potomac basin, in nearly the same line, and the coal-fields of the Little Sewell, and the E. side of the Big Sewell Mountain in Virginia. The position of this belt of the Coal-measures is somewhat W. of the region of steep flexures of the strata, and beyond all the considerable dislocations, while it embraces a few very extensive, regular, and nearly symmetrical anticlinal axes of the flatter form, distinctive of their intermediate position between the E. and W.

Thirdly. The great Appalachian basin, with its subordinate troughs, forming the wide coal-field watered by the Ohio River and its tributaries, embraces a series of coal-beds, which are all distinguished by a still larger amount of volatile matter. In crossing the breadth of this wide coal-field, we find a very material alteration in the character and composition of the coal. Along its E. side, or near the last considerable axis of the Appalachian chain, the amount of volatile matter is commonly from thirty to thirty-five per cent. Westward of this line, on the Monongahela River, both in Pennsylvania and Virginia, the proportion approaches to forty per cent.; while still farther in the same direction, or near the Ohio River, it ranges from forty to even fifty per cent., according to local circumstances. In this most Western or main coal-field, the flexures of the strata are extremely gentle, and comparatively wide apart; but even here we observe a beautiful progression in the amount of the bitumen, as we recede from the very low axes which traverse the S. E. side of the great plain. What renders the foregoing comparison of the several ranges of the coal-formation particularly exact and satisfactory, is the circumstance that, in more than one instance, we are enabled to trace the very same coal-seam through its various degrees of bituminization, from an almost true anthracite to a form in which it possesses a full proportion of volatile matter. Thus the great Pittsburg Bed, to take it as an example, contains on the Potomac, in some localities, as little as 15.5 per cent.; but near the Eastern margin of the great Western basin, as at Blairsville, and again in Virginia, it has about thirty-one per cent.; and towards the middle of the main basin at Pittsburg and on the Kenawha, as much as from forty to forty-three per cent.

The cause of the different degrees of de-bituminization of the coals, in different parts of their range, I am disposed to attribute to the prodigious quantity of intensely-heated steam and gaseous matter emitted through the crust of the earth, by the almost infinite number of cracks and crevices which must have been produced during the undulation and permanent bending of the strata. All the phenomena of modern earthquakes and volcanoes warrant us in supposing that the elevation of our coal-rocks, if effected in the manner I have imagined, must have been accompanied by the escape of an immense amount of hot vapors, the chemical and thermal agency of which cannot be overlooked, upon any hypothesis of the rending and uplifting of great mountain-tracts. It is easy to conceive that the coal, throughout all the E. basins, if thus effectually steamed and raised in temperature in every part of its mass, would discharge a greater or less proportion of its bitumen and other volatile constituents, as the strata were more or less frequently and violantly undulated by earthquake action. It is also obvious that the more Western beds, remoter from the region of active movements, less crushed and fissured, and presenting a greater resistance to permeation by the subterranean vapors, would, in virtue of their mere geographical position in the chain, be much less extensively de-bituminized. The striking fact that we nowhere, not even in the most dislocated and disturbed districts of the anthracite coal-field, find any traces of true igneous rocks, that, by their contiguity to the coal, could have caused the loss of its bitumen, is a circumstance in their geology which goes far to confirm the truth of the hypothesis. Precisely in proportion as the flexures of the strata diminish in our progress W., does the quantity of the bitumen in the coal augment; but it is difficult to conceive how any such law of gradation could have been the result of a temperature transmitted by conduction from the general lava mass beneath the crust, for that would imply a corresponding increasing gradation in the thickness of the crust, advancing W. under the coal-fields; whereas such an inference is in direct conflict with the fact of the general diminution W. of the Appalachian rocks, besides being inconsistent with all correct geothermal considerations, which forbid our imagining so unequal a conduction, to the surface, of the earth's interior temperature.

INDEX

To Report of Progress in Clearfield and Jefferson Counties, Pa., 1874, *by Franklin Platt, Assistant Geologist.*

APPENDIX TO F PLATT'S REPORT OF PROGRESS, 1874.

Levels, above tide, of points on the "Tyrone and Clearfield Rail road," in Clearfield county.

2,031 Emigh's Summit.
1,897 Sandy Ridge.
1,784 Powelton.
1,471 Osceola.
1,443 Dunbar.
1,411 Philipsburg.
1,728 Turner's Summit.
1,737 Ross's Summit.
1,413 Woodland Summit.
1,126 Clearfield Creek Bridge.
1,089 Clearfield.
1,111 Susquehanna Bridge.
1,127 Curwinsville.
1,179 Bridgeport.
1 513 Blue Ball: 24th mile post.
1,675 Wallacetown: 27th mile post.
1,655 Bigler: 30th mile post.

Levels, above tide, of stations on the Bennet Branch Extension of the Allegheny Valley Railroad.

788,084 Philadelphia and Erie Railroad Junction.
848,195 Mix Run Station.
880,000 Miller's Station.
897,820 Dent's Run Station.
938,480 Enz Station.
949,000 Grant Station.
973,000 Mount Pleasant Station.
993,000 Devil's Elbow Station.
1,014,000 Benezette Station.
1,121,580 Caledonia Tunnel (250' E.)
1,124,000 Caledonia Tunnel (370' W.)
1,163,400 Slabtown Dam.
1,439,930 Summit Tunnel, (east end.)
1,436,610 Summit Tunnel, (west end.)
1,374,110 Evergreen Water Station.
1,361,000 Maghee's.
1,350,960 Reynoldsville.
1,335,000 Prindible's.
1,300,650 Fuller's Mill Station.
1,272,812 Iowa Mill Station.
1,255,820 Gooseneck Station.
1,239,940 Bell's Mills.
1,234,829 Garrison and Fuller's Mill.
1,209,430 Brookeville Station.
1,199,000 Nicholson's Mill.
1,189,380 Puckerty Point.
1,161,000 Troy Station.
1,136,710 Heathville.
1,082,020 Maysville.
1,066,500 Millville.
1,063,000 Indiantown Run.
1,054,000 New Bethlehem.
1,024,500 Anthony's Neck Tunnel, (W.)
893,000 Lawsonham.
824,700 Main Line Intersection.

Levels, above tide, of various mines in Clearfield county, mentioned in this report.

1,528 Franklin Mine.
1,530 Penn Mine.
1,517 Eureka Mine.
1,487 Stirling Mine.
1,546 Beaverton Mine.
1,503 Moshannon Mine.
1,515 Decatur Mine.
1,494 Clearfield Co.'s Mine.
1,443 S. W. Osceola Dam.
1,543 Logan Bank.
1,584 End of Mapleton Br. Railroad.
1,531 Mapleton Mine.
1,442 Decatur Mine.
1,476 Morrisdale Mine.
1,439 Bell's Mine.

Levels, above tide, of various mines in Jefferson county, mentioned in this report.

1,485 Hoover's Mine.
1,479 Prescott's Mine.
1,489 Carrier and Wilson's Mine.
1,504 Seley's Mine.
1,516 Spragues' Mine.
1,472 M'Creight's Mine.
1,501 Strouse's Mine.
1,491 Sheesley's Mine.
1,490 Wood Reynolds' Mine.
1,524 Brown & Sharp's Mine.
1,417 Dixon's Mine.
1,472 Rumbarger's Mine.
1,442 Rumbarger's Mine.
1,515 Pancoast's Mine.
1,613 Phillipi's Mine.
1,554 Syphrit's Mine.
1,469 Sheafer's Mine.
1,658 Secrit's Mine.
1,744 Norris's Mine.
1,807 Brown's Mine.
1,652 Uplinger's Mine.
1,488 Zeitler's Mine.
1,281 Hawk's (P.) Mine.
1,286 Carmalt's Mine.
1,376 Hum Mine.
1,397 Morris's (J. B.) Mine, (west.)
1,294 Morris's (J. B.) Mine, east
1,264 Jone's Mine.
1,588 Pantall's Mine.
1,498 M'Kee's Mine.
1,486 Smith's Mine.
1,451 Wengert's Mine.
1,552 Hilles's Mine.
1,306 Thomas's (S.) Mine.
1,380 Weaver's (Christian) Mine.
1,373 Conrad's Mine.
1,570 Mean's (Thomas) Mine.
1,571 Ruth's Mine.
1,551 Anthony's Mine.
1,503 Stewart's Mine.
1,550 Cox's (P.) Mine.
1,579 Crawford's (S.) Mine.
1,660 Dougherty's Lower Mine.
1,701 Dougherty's Upper Mine.
1,636 Smith's (A. H.) Mine
1,624 Stewart's (J. J.) Mine.
1,716 Denneson's (David) Mine.
1,560 Meeker's Mine.
1,613 Morrison's (S.) Mine.
1,524 Frost's Mine.
1,684 M'Donald's (R.) Mine.
1,622 M'Donald's (R.) Mine.

Other Levels.

1,433 Brockwayville.
1,193 S. W. Mahoning creek, mouth of Elk run.
1,353 Reynoldsville Station. Railroad track
1,566 Rockdale.
1,547 Rockdale.
1,206 Punxsutawney.

www.ingramcontent.com/pod-product-compliance
Lightning Source LLC
LaVergne TN
LVHW020239110826
845151LV00003B/974